Switched-Inductor

Power Supplies

By

Gabriel Alfonso Rincón-Mora

School of Electrical and Computer Engineering
Georgia Institute of Technology

Rincon-Mora.gatech.edu

Contents

List of Figures

List of Abbreviations

BJT ≡ Bipolar-Junction Transistor

FET ≡ Field-Effect Transistor

LED ≡ Light-Emitting Diode

MOS ≡ Metal–Oxide–Semiconductor

RMS ≡ Root–Mean–Squared

CCM ≡ Continuous-Conduction Mode

DCM ≡ Discontinuous-Conduction Mode

A_{Si} ≡ Silicon Area

d_D ≡ Drain Duty Cycle

d_{DO} ≡ Output Duty Cycle

d_E ≡ Energize Duty Cycle

d_{EI} ≡ Input Duty Cycle

E_M, E_L ≡ Magnetic Energy

f_{LC} ≡ LC Resonant Frequency

f_{SW} ≡ Switching Frequency

i_{IN} ≡ Input Current

i_L ≡ Inductor Current

$i_{L(HI)}$ ≡ Inductor Current's High CCM Peak

$i_{L(LO)}$ ≡ Inductor Current's Low CCM Peak

$i_{L(MIN)}$ ≡ Inductor Current's Minimum

$i_{L(PK)}$ ≡ Inductor Current's DCM Peak

i_O ≡ Output Current

Δi_L ≡ Inductor Ripple Current

Δi_{LD} ≡ Load Dump

$k_C \equiv$ Coupling Factor/Coefficient

$k_L \equiv$ Transformer Translation

$P_{IN} \equiv$ Input Power

$P_{LOSS} \equiv$ Power Losses

$P_O \equiv$ Output Power

$P_R \equiv$ Ohmic Power

$R_{CH} \equiv$ MOS Channel Resistance

$R_{ESR} \equiv$ Equivalent Series Resistance

$R_{SER} \equiv$ Series Resistance

$t_C \equiv$ Conduction Time

$t_D \equiv$ Drain Time

$t_{DT} \equiv$ Dead Time

$t_E \equiv$ Energize Time

$t_{SW} \equiv$ Switching Period

$\tau_{LC} \equiv$ LC Resonant Period

$\sigma_{LOSS} \equiv$ Fractional Losses

$v_D \equiv$ Drain Voltage

$v_E \equiv$ Energize Voltage

$v_G \equiv$ Gate Voltage

$v_{IN} \equiv$ Input Node/Voltage

$v_L \equiv$ Inductor Voltage

$v_O \equiv$ Output Node/Voltage

$v_S \equiv$ Source Voltage

$v_{SWI} \equiv$ Switching Input Node/Voltage

$v_{SWO} \equiv$ Switching Output Node/Voltage

$v_T \equiv$ Threshold Voltage

$\Delta v_O \equiv$ Output Ripple Voltage

Switched-Inductor Power Supplies

The fundamental purpose of power supplies is to transfer power. *Switched inductors* (SLs) are pervasive in power-supply systems because they output a large fraction of the power they draw from an input source. The fundamental reason for this is low Ohmic losses, and that's because switches in the network only drop millivolts. So the *Ohmic power* P_R that these switches burn when they conduct current i_{SW} across these low voltages v_{SW} is low at $i_{SW}v_{SW}$. This means that *power-conversion efficiency* η_C, which is the fraction of *input power* P_{IN} delivered as *output power* P_O that *power losses* P_{LOSS} avail, is usually high between 85% and 95% when P_{IN} is moderate to high:

$$\eta_C \equiv \frac{P_O}{P_{IN}} = \frac{P_{IN} - P_{LOSS}}{P_{IN}} = 1 - \frac{P_{LOSS}}{P_{IN}} = 1 - \sigma_{LOSS}. \qquad (1)$$

Fractional losses σ_{LOSS} or P_{LOSS}/P_{IN} in the system therefore limit efficiency.

1. Transfer Media

1.1. Inductor

A. Ideal Inductor

Inductors magnetize and de-magnetize with voltages of opposing polarity. In other words, they energize and drain their magnetic fields with positive and negative voltages. So as the inverting *inductor voltage* $\pm v_L$ in Fig. 1 raises and lowers *inductor current* i_L across time t_X, the *magnetic energy* E_M or E_L that the *transfer inductor* L_X holds rises and falls with the square of i_L:

$$i_L = \left(\frac{v_L}{L_X}\right) t_X \qquad (2)$$

and
$$E_M \equiv E_L = 0.5 L_X i_L^{\,2}.$$
(3)

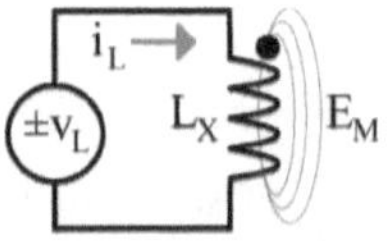

Fig. 1. Magnetizing inductor.

$+v_L$ is the *energize voltage* $+v_E$ that magnetizes L_X and $-v_L$ is the *drain voltage* $-v_D$ that de-magnetizes L_X. But since i_L can also flow in the opposite direction, $-v_L$ can be the $+v_E$ that energizes L_X and $+v_L$ the $-v_D$ that drains L_X. Note that $+v_L$ supplies power when i_L in Fig. 1 flows to the right and sinks power when i_L flows to the left and $-v_L$ similarly supplies power when i_L flows to the left and sinks power when i_L flows to the right.

L_X's inductance is basically a measure of how much energy i_L can hold. That is to say, a higher L_X holds more energy with the same current. Since more magnetic space stores more E_M, L_X scales with the *cross-sectional area* A_L of the coil used to build the inductor and the *number of turns* N_L in the coil. L_X scales more with N_L when the loops align because the magnetic fields of the loops reinforce one another to establish an even stronger field.

B. Actual Inductor

In practice, the coil is resistive, so inductors burn Ohmic power. *Equivalent series resistance* R_{ESR} scales with N_L because the length of the resulting wire is longer with more turns. Unfortunately, nearby coils exert a field that pushes current away from the edges. So R_{ESR} also scales with the number of nearby coils. This is the *proximity effect*.

Fast-moving charges also tend to flow along the outer edges of the coil. This is why R_{ESR} also scales with frequency f_L. This is the *skin effect*. So from a circuit's perspective, P_R in R_{ESR} decomposes into the low- and high-frequency components that dc and ac resistances $R_{ESR(DC)}$ and $R_{ESR(AC)}$ consume:

$$P_{ESR} = i_{L(AVG)}^{\ 2} R_{ESR(DC)} + \Delta i_{L(RMS)}^{\ 2} R_{ESR(AC)}. \tag{4}$$

Here, $i_{L(AVG)}$ is i_L's average dc current and $\Delta i_{L(RMS)}$ is the *root–mean–squared* (RMS) equivalent of the ac ripple component.

C. Optimal Inductor

The optimal L_X carries lots of E_L and burns little P_{ESR}. L_X should therefore deliver high E_L with low i_L. This happens when *magnetic permeability*, which is the ability to form a magnetic field, is high, for which a large magnetic core is necessary. R_{ESR} should also be low, so the coil should be thick. Although higher N_L raises L_X, lengthening the coil and the proximity effect of more nearby turns also raise R_{ESR}. In other words, the optimal L_X is, unfortunately, large. So confining L_X to smaller dimensions and cheaper materials sacrifices power for space and cost.

1.2. Transformer

A. Ideal Transformer

A transformer is nothing but two inductor coils that share the same magnetic space and have access to the same magnetic field. So when neglecting unintended parasitic effects, L_I and L_O in Fig. 2 hold the same E_M:

$$E_M = 0.5 L_I i_{LI}^{\ 2} = 0.5 L_O i_{LO}^{\ 2}. \tag{5}$$

This means that the coil with the higher inductance conducts less current. In other words, L_I's current i_{LI} is the k_L translation of L_O's current i_{LO} that L_O/L_I establish, which means i_{LI} is a k_L fraction of i_{LO}:

$$\frac{i_{LI}}{i_{LO}} = \sqrt{\frac{L_O}{L_I}} \equiv k_L. \tag{6}$$

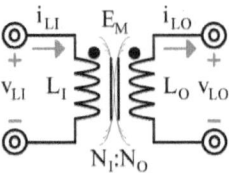

Fig. 2. Ideal transformer.

The ideal transformer incorporates zero resistance. As a result, P_R is negligible, so the output receives with P_O what that the input supplies with P_{IN}:

$$P_{IN} = i_{LI} v_{LI} = P_O = i_{LO} v_{LO}. \tag{7}$$

L_O's voltage v_{LO} is therefore the k_L translation that i_{LI}/i_{LO}, and ultimately, L_O/L_I dictate. In other words, v_{LO} is a k_L fraction of v_{LI}:

$$k_C \equiv \frac{v_{LO}}{v_{LI}} = \frac{i_{LI}}{i_{LO}} = k_L = \sqrt{\frac{L_O}{L_I}}. \tag{8}$$

This v_{LI}–v_{LO} translation is what the *coupling factor* or *coupling coefficient* k_C describes. When L_I's and L_O's geometries match, k_L reduces to the ratio of the number of loops in L_O to those in L_I. This is the *turns ratio* to which literature refers when using N_O/N_I to describe k_L.

Example 1: Derive an expression for P_{IN} when v_{IN} supplies L_I and R_{LD} loads L_O.

Solution:

$$P_{IN} = i_{LI} v_{IN} = i_{LO} k_L v_{IN}$$

$$= \left(\frac{v_{LO}}{R_{LD}}\right) k_L v_{IN} = \left(\frac{v_{IN} k_L}{R_{LD}}\right) k_L v_{IN}$$

$$= \frac{\left(v_{IN} k_L\right)^2}{R_{LD}} = P_{LD} = P_O$$

B. Actual Transformer

In practice, coupled inductors access a fraction of the magnetic field they share. So when decomposed into the pieces that actually couple: L_I and L_O, and the ones that do not: L_I' and L_O', like Fig. 3 illustrates, only a k_{CI} fraction of the input couples to a k_{CO} fraction of the output:

$$k_{CI} = \frac{L_I}{L_I + L_I'} \tag{9}$$

and

$$k_{CO} = \frac{L_O}{L_O + L_O'}. \tag{10}$$

This means that L_O avails a k_{CI} fraction of the energy that v_{LI} supplies, so L_I' reduces k_C to

$$k_C \equiv \frac{v_{LO}}{v_{LI}} = \frac{v_{LI}k_{CI}k_L}{v_{LI}} = \left(\frac{L_I}{L_I + L_I'}\right)\sqrt{\frac{L_O}{L_I}}. \tag{11}$$

This also means that L_O' is a load to L_O.

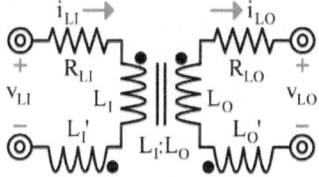

Fig. 3. Actual transformer.

Since separating the coils decreases the fraction of inductance that couples, L_I and L_O and their resulting k_C fall with increasing separation d_X. Misalignment between the coils also reduces k_C. Coupling also depends on the geometry of the coils, so variations in d_X manifest in k_C in a variety of ways.

L_I and L_O are also resistive, and as a result, not lossless. So input and output resistances R_{LI} and R_{LO} further alter the k_C fraction that couples and reaches v_{LO}. Needless to say, better transformers *couple* more and *resist* less.

2. Switched Inductors

Switched inductors energize and drain in alternating phases of a cycle. An input source at v_{IN} first energizes a transfer inductor L_X (in Fig. 4) across an *energize time* t_E. Note that the energize voltage v_E should always include elements of the *input voltage* v_{IN}. The load that the output v_O feeds then drains L_X across a *drain time* t_D. For this, the switching network connects L_X in such a way that elements of the *output voltage* v_O apply an opposing v_D across L_X. This way, L_X drains into v_O.

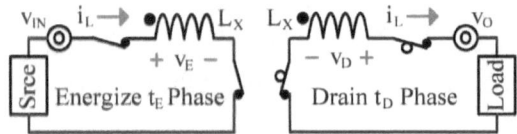

Fig. 4. Phases of the switched inductor.

2.1. Static Power-Supply Applications

Many consumer applications transfer power from static dc sources to loads that impose or require steady voltages. Conventional sources include *lithium-ion* (Li-Ion), *nickel* (Ni), and lead-acid batteries and ac–dc rectifiers that convert dynamic ac sources to static dc outputs. Chargers recharge Li-Ion and nickel batteries, voltage regulators feed systems that require steady supplies, and *light-emitting diode* (LED) drivers feed steady diode voltages. These outputs are essentially dc because the capacity (or equivalent capacitance) of the battery is very high and the feedback controllers in the regulator and LED driver keep v_O and i_O steady. Although controllers cannot respond instantaneously, designers normally add capacitance to their outputs. So SL inputs and outputs in all these types of applications are nearly dc.

2.2. Inductor Current

Since v_{IN} and v_O in dc–dc applications are static, the i_L that v_L's v_E and v_D establish is a linear ramp di_L/dt_X:

$$\frac{di_L}{dt_X} = \frac{v_L}{L_X}. \tag{12}$$

So like Fig. 5 shows, v_E ramps i_L up linearly across t_E. The opposing voltage that v_D applies similarly ramps i_L down across t_D. t_E and t_D are opposite phases of the *conduction time* t_C. Energizing and draining L_X this way across t_C produces the triangular *current ripple* Δi_L shown.

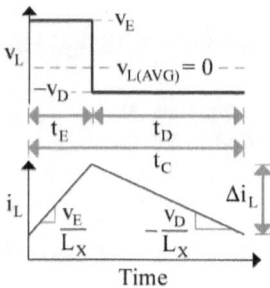

Fig. 5. Inductor waveforms in dc-supplied switched inductors.

2.3. Duty Cycle

The *energize duty cycle* d_E is the t_E fraction of t_C across which v_E energizes L_X:

$$d_E \equiv \frac{t_E}{t_C}. \tag{13}$$

The *drain duty cycle* d_D is the opposite: the t_D fraction of t_C across which v_D drains L_X:

$$d_D \equiv \frac{t_D}{t_C} = \frac{t_C - t_E}{t_C} = 1 - d_E. \tag{14}$$

But since t_D is the time that remains across t_C after t_E elapses, t_D is also $t_C - t_E$ and d_D is, in consequence, $1 - d_E$.

Under steady-state conditions, i_L should rise in one cycle as much as i_L falls. v_E across t_E therefore raises i_L by the same Δi_L that v_D across t_D lowers i_L:

$$\Delta i_L = \left(\frac{v_E}{L_X}\right)t_E = \left(\frac{v_D}{L_X}\right)t_D. \tag{15}$$

This means that time, duty-cycle, and reciprocal voltage ratios t_E/t_D, d_E/d_D, and v_D/v_E match:

$$\frac{t_E}{t_D} = \frac{d_E}{d_D} = \frac{d_E}{1-d_E} = \frac{v_D}{v_E}, \tag{16}$$

and d_E is a v_D fraction of the voltages v_E and v_D:

$$d_E = \frac{v_D}{v_E + v_D}. \tag{17}$$

A. Ohmic Losses

A lossless L_X delivers all the energy that it receives from v_{IN}. In practice, however, R_{ESR} burns energy E_R that L_X does not receive or deliver. R_{ESR} also drops a voltage v_R that reduces v_E and v_D. Reducing E_L and v_E extends the t_E that L_X needs to energize across Δi_L in Fig. 6.

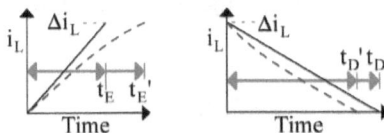

Fig. 6. Inductor current with Ohmic losses.

Reducing v_D similarly extends t_D. E_R counters this effect because E_R drains L_X. Since v_R scales linearly with i_L and E_R's $i_{L(RMS)}^2 R_{ESR}$ scales quadratically, E_R shortens t_D more than v_R extends t_D when i_L is high. In all, i_L is increasingly less linear (and more parabolic) with higher R_{ESR}.

These t_E and t_D variations alter d_E. When v_D is greater than v_E, v_R is a smaller fraction of v_D than $2v_R$ is of $v_E + v_D$, so the resulting d_E' is higher than d_E. The opposite is true when v_E is greater than v_D:

$$d_E' = \frac{v_D - v_{RL(AVG)}}{\left(v_E - v_{RL(AVG)}\right) + \left(v_D - v_{RL(AVG)}\right)} = \frac{v_D - i_{L(AVG)}R_{ESR}}{v_E + v_D - 2i_{L(AVG)}R_{ESR}}. \tag{18}$$

2.4. Continuous Conduction

In *continuous-conduction mode* (CCM), L_X conducts continuously. In CCM, L_X's conduction period extends across the entire *switching period* t_{SW}. In other words, i_L is never static – di_L/dt is never 0. In steady state, i_L's *low* and *high CCM peaks* $i_{L(LO)}$ and $i_{L(HI)}$ do not vary, so i_L ripples periodically like Fig. 7 shows. Since t_C is t_{SW} in CCM, d_E and d_D become

$$d_E\Big|_{CCM} = \frac{t_E}{t_C}\Big|_{CCM} = \frac{t_E}{t_{SW}} \tag{19}$$

and

$$d_D\Big|_{CCM} = \frac{t_D}{t_C}\Big|_{CCM} = \frac{t_D}{t_{SW}} = 1 - d_E. \tag{20}$$

Note that i_L ripples about i_L's average $i_{L(AVG)}$, $i_{L(AVG)}$ is i_L's low plus half i_L's ripple, and $i_{L(AVG)}$ can be positive or negative, which is another way of saying i_L can reverse direction:

$$i_{L(AVG)}\Big|_{CCM} = i_{L(LO)} + \Delta i_{L(AVG)}\Big|_{CCM} = i_{L(LO)} + 0.5\Delta i_L. \tag{21}$$

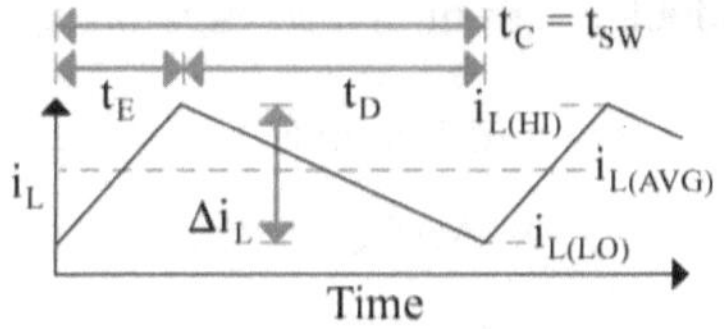

Fig. 7. Inductor current in continuous conduction.

Example 2: Determine t_E, t_D, and Δi_L in CCM when v_E is 2 V, v_D is 1 V, t_{SW} is 1 μs, and L_X is 10 μH.

Solution:

$$d_E = \frac{v_D}{v_E + v_D} = \frac{1}{2+1} = 33\%$$

$$\therefore \quad t_E = d_E t_C = d_E t_{SW} = 330 \ ns$$

$$t_D = t_C - t_E = t_{SW} - t_E = 670 \ ns$$

$$\Delta i_L = \left(\frac{v_E}{L_X}\right) t_E = \left(\frac{2}{10\mu}\right)(330n) = 66 \ \text{mA}$$

2.5. Discontinuous Conduction

In *discontinuous-conduction mode* (DCM), L_X conducts discontinuously. To be more specific, L_X's conduction does not extend across t_{SW}, like Fig. 8 shows. So L_X first energizes and drains and then remains static until the next cycle begins, which means di_L/dt is greater than, less than, and equal to zero. Since t_C is not t_{SW}, d_E and d_D do not relate to t_{SW} like they do in CCM. d_E and d_D should therefore remain in their more general forms: t_E/t_C and t_D/t_C. Note that, for i_L to be static (i.e., remain unchanged), L_X's v_L must be practically zero.

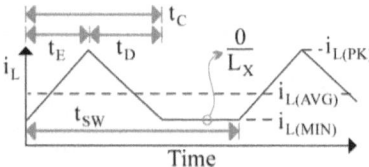

Fig. 8. Inductor current in discontinuous conduction.

DCM is ultimately possible because the system does not allow i_L to fall below i_L's *minimum* $i_{L(MIN)}$. And even then, DCM occurs only when i_L reaches $i_{L(MIN)}$ before t_{SW} ends. Generally, $i_{L(AVG)}$ is $i_{L(LO)}$ plus Δi_L's average, but since i_L reaches $i_{L(MIN)}$ before t_{SW} ends in DCM, $i_{L(AVG)}$ is $i_{L(MIN)}$ plus less than half i_L's *DCM peak* $i_{L(PK)}$:

$$i_{L(AVG)}\big|_{DCM} = i_{L(MIN)} + \Delta i_{L(AVG)}\big|_{DCM} = i_{L(MIN)} + \left(\frac{i_{L(PK)}}{2}\right)\left(\frac{t_C}{t_{SW}}\right). \qquad (22)$$

Note that $i_{L(AVG)}$ is $i_{L(MIN)}$ plus half $i_{L(PK)}$ when t_C is t_{SW}, which corresponds to CCM's $i_{L(AVG)}$, which is $i_{L(LO)} + 0.5\Delta i_L$. This means that L_X enters DCM when $i_{L(AVG)}$ falls below $i_{L(MIN)}$ plus half CCM's ripple Δi_L.

Engineers often think of $i_{L(MIN)}$ being zero in DCM, in which case L_X energizes and depletes before the termination of the cycle. Some engineers use *pseudo discontinuous-conduction mode* (PDCM) to refer to cases where $i_{L(MIN)}$ is not zero. Technically, PDCM is a special case of the more general DCM category. Although most often applied to zero cases, $i_{L(MIN)}$ in DCM can be any value.

Example 3: Determine t_E, t_D, and Δi_L when v_E is 2 V, v_D is 1 V, $i_{L(MIN)}$ is zero, $i_{L(AVG)}$ is 25 mA, t_{SW} is 1 µs, and L_X is 10 µH.

Solution:

$d_E = 33\%$ and $\Delta i_L = 66$ mA in CCM from previous example.

$$i_{L(AVG)} = 25 \text{ mA} < i_{L(MIN)} + 0.5\Delta i_L\big|_{CCM}$$

$$= 0 + 0.5(66m) = 33 \text{ mA}$$

$\therefore$ DCM

$$i_{L(AVG)} = i_{L(MIN)} + \left(\frac{i_{L(PK)}}{2}\right)\left(\frac{t_C}{t_{SW}}\right) = 0 + \left(\frac{i_{L(PK)}}{2}\right)\left(\frac{t_C}{t_{SW}}\right)$$

$$i_{L(PK)} = \left(\frac{v_E}{L_X}\right)t_E = \left(\frac{v_E}{L_X}\right)d_E t_C$$

$\rightarrow$

$$t_C = \sqrt{2i_{L(AVG)}\left(\frac{L_X}{v_E}\right)\left(\frac{t_{SW}}{d_E}\right)}$$

$$= \sqrt{2(25m)\left(\frac{10\mu}{2}\right)\left(\frac{1\mu}{0.33}\right)} = 870 \text{ ns}$$

$$t_E = d_E t_C = (0.33)(870n) = 290 \text{ ns}$$

$$t_D = t_C - t_E = 870n - 290n = 580 \text{ ns}$$

$$i_{L(PK)} = \left(\frac{v_E}{L_X}\right)t_E = \left(\frac{2}{10\mu}\right)(290n) = 58 \text{ mA}$$

2.6. CMOS Implementations

Switches in a *complementary metal–oxide–semiconductor* (CMOS) implementation are *MOS field-effect transistors* (MOSFETs). Replacing each switch with the parallel combination of complementary N- and P-channel MOSFETs is the most straightforward translation, though not always the most effective one. Available *gate drive* v_{GST} or $v_{GS} - v_T$ and the *sheet resistivity* R_{SH} that v_{GST} engenders can dictate which type of transistor is more efficient. This is why switches in many applications are N- *or* P-channel transistors, not the parallel combination of both.

Although *electron mobility* μ_N is usually two to three times greater than *hole mobility* μ_P, v_{GST} can trump that difference in R_{SH}:

$$R_{SH} = R_{CH}\left(\frac{W}{L}\right) = \left.\frac{v_{DS}}{i_{D(TRI)}}\left(\frac{W}{L}\right)\right|_{v_{DS}\ll v_{DS(SAT)}} \approx \frac{1}{C_{OX}{}''\mu_{N/P}\left|v_{GST}\right|}, \qquad (23)$$

where R_{CH} is *channel resistance*, W and L are the width and length of the channel, and $i_{D(TRI)}$ is drain current in *triode* when v_{DS} is much lower than the saturation level $V_{DS(SAT)}$. Still, NFETs are less resistive than PFETs under equivalent v_{GST}'s and *silicon areas* A_{Si}. A PFET is therefore less resistive only when its v_{SGT} is more than μ_N/μ_P times (or 2× to 3×) greater than the v_{GST} of an NFET.

The *source voltage* v_S is key to determining v_{GS} in v_{GST}. This v_S is also the voltage that the *drain voltage* v_D approaches (within 10–200 mV) after the MOSFET closes. v_{GS} is usually higher than v_{SG} when v_S is low because a higher *gate voltage* v_G is more likely than a lower v_G. v_{SGT} is similarly higher than v_{GST} when v_S is high because a lower v_G is more likely than a higher v_G. So generally, NFETs are good *low-side switches* and PFETs are good *high-side switches*. When v_S is neither high nor low, *threshold voltage* v_T plays a more decisive role in v_{GST} and v_{SGT}.

After using v_{GST} and v_{SGT} to determine which FET is least resistive, the bulk terminal is next. To start, v_B should not "float" because disconnected nodes are vulnerable to noise, and in the case of v_B, v_T is sensitive to v_B. But connecting v_B to v_S will expose v_D's body diode and connecting v_B to v_D will expose v_S's body diode. So the connection should short the body diode that conducts when not intended.

Energizing switches are usually transistors because L_X should energize only when prompted. Since L_X already conducts current when energized, *asynchronous* realizations use diodes to steer and drain i_L into v_O. The advantage of this approach is that diodes do not require a synchronizing control signal to switch. *Synchronous* power supplies, however, use transistors only, so they need a synchronizing signal to close and open all MOSFETs, including the ones that drain L_X.

Since energizing events should be synchronized to commands, body diodes should not energize L_X asynchronously. So the body connections of energizing switches should short the body diodes that would conduct *input current* i_{IN}. Drain switches, on the other hand, should block reverse *output current* i_O. Their body connections should therefore short the body diodes that would conduct reverse i_O.

Synchronous converters are popular in many applications because diodes can consume more power with the 0.60–0.75 V that they drop than MOSFETs can with the 10–200 mV that MOSFETs drop. But since gate signals can easily crisscross, adjacent MOSFETs can inadvertently short their inputs together. Controllers must therefore insert *dead time* t_{DT} between the conduction times of adjacent switches. Ironically, diodes must conduct L_X's current across these dead times because all MOSFETs are off across that time.

2.7. Design Limits

i_L is linear across t_E and t_D when P_R is low. This happens when *series resistances* R_{SER} drop a small fraction of v_E and v_D. So the $v_{R(MAX)}$ that $i_{L(HI)}$ into R_{SER} produces should be less than a fraction of the lowest v_L:

$$V_{R(MAX)} = i_{L(HI)} R_{SER} = \left(i_{L(AVG)} + 0.5\Delta i_L \right) R_{SER} \leq 10\% v_{L(MIN)}. \qquad (24)$$

Since R_{SER} includes the resistances of one or two switches and L_X, this $v_{R(MAX)}$ limits the switches' R_{CH}'s and the maximum R_{ESR} allowed.

Feeding L_X's triangular i_L into C_O produces an *output ripple voltage* Δv_O that also deviates v_O in v_D from its intended setting. C_O should therefore be high enough to keep Δv_O within a small fraction of v_O:

$$\Delta v_{O(MAX)} \leq 10\% v_O. \qquad (25)$$

C_O in voltage regulators is on the order of microfarads primarily for this reason: to keep i_L and wide and sudden *load dumps* $\Delta i_{LD(MAX)}$ from deviating v_O beyond this $\Delta v_{O(MAX)}$ limit. C_O for batteries is so high (on the order of millifarads) that Δv_O in chargers is normally very low.

When interconnected, inductors and capacitors exchange energy every quarter cycle of their *resonant period* τ_{LC}:

$$\tau_{LC} = 2\pi \sqrt{L_X C_O}. \qquad (26)$$

So their voltages and currents oscillate at the *resonant frequency* f_{LC} that $1/\tau_{LC}$ sets. Since L_X in switched inductors drains into v_O, L_X connects to C_O no less than a t_D fraction of t_{SW}. If τ_{LC} is shorter than t_{SW}, L_X and C_O can exchange energy a few times before t_{SW} ends. To avoid these oscillations, t_{SW} should be much shorter than τ_{LC}'s quarter cycle:

$$t_{SW} \leq 10\% \left(\frac{\tau_{LC}}{4} \right) = 2.5\% \tau_{LC}, \qquad (27)$$

which means the *switching frequency* f_{SW} should be correspondingly higher than f_{LC}.

Example 4: Determine R_{CH} and f_{SW} limits for two switches that conduct i_L when v_D is 1 V, L_X is 10 μH, R_L is 200 mΩ, C_O is 10 μF, $i_{L(AVG)}$ is 70 mA, and Δi_L is 30 mA.

Solution:

$$R_{SER} \leq \frac{10\%v_D}{i_{L(HI)}} = \frac{10\%v_D}{i_{L(AVG)}+0.5\Delta i_L} = \frac{10\%(1)}{70m+0.5(30m)} = 1.2 \ \Omega$$

$$\rightarrow \quad R_{CH} \leq \frac{R_{SER}-R_L}{2} = \frac{1.2-200m}{2} = 500 \ m\Omega$$

$$t_{SW} \leq 2.5\%\tau_{LC} = 2.5\%(2\pi)\sqrt{L_X C_O}$$

$$= (0.0025)(2\pi)\sqrt{(10\mu)(10\mu)} = 1.6 \ \mu s$$

$$\rightarrow \quad f_{SW} \geq 640 \text{ kHz}$$

3. Buck–Boost

3.1. Ideal Buck–Boost

A. Power Stage

As the name implies, buck–boost SLs *buck* or *boost* v_{IN} to a *lower* or *higher* v_O. Energize switches S_{EI} and S_{EG} in the buck–boost of Fig. 9 draw input power by connecting L_X across v_{IN} and ground. This way, with a positive v_E that is equal to v_{IN}, L_X energizes.

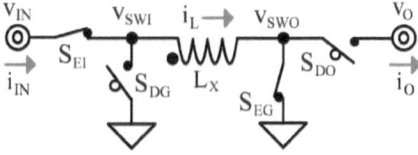

Fig. 9. Ideal buck–boost.

When done energizing, S_{EI} and S_{EG} open and drain switches S_{DG} and S_{DO} close to connect L_X to v_O. v_L's polarity reverses this way to $-v_O$, so

L_X drains into v_O. v_L in Fig. 5 therefore swings between $+v_E$'s v_{IN} and $-v_D$'s $-v_O$ and i_L ramps up with $+v_E$'s v_{IN} and down with $-v_D$'s $-v_O$.

B. Duty-Cycle Translation

In steady state, v_L's average is zero, which is why engineers often say inductors are "dc shorts". The switching voltages v_{SWI} and v_{SWO} in Fig. 9 are therefore, on average, equal. Since v_{IN} connects to v_{SWI} a t_E fraction of t_C and v_O connects to v_{SWO} a t_D fraction of t_C, their averages are matching duty-cycled fractions of v_{IN} and v_O:

$$
V_{SW(AVG)} = V_{SWI(AVG)} = V_E\left(\frac{t_E}{t_C}\right) = V_{IN}d_E
$$

$$
= V_{SWO(AVG)} = V_D\left(\frac{t_D}{t_C}\right) = v_O d_D = v_O\left(1 - d_E\right)
$$

(28)

This means that v_O is a duty-cycled scalar d_E/d_D of v_{IN}:

$$
V_O = V_{IN}\left(\frac{d_E}{d_D}\right) = V_{IN}\left(\frac{d_E}{1 - d_E}\right).
$$

(29)

And since d_D is $1 - d_E$, d_E is a v_O fraction of $v_{IN} + v_O$:

$$
d_E = \frac{V_D}{V_E + V_D} = \frac{V_O}{V_{IN} + V_O}.
$$

(30)

When d_E is greater than 50%, the average switching voltage $v_{SW(AVG)}$ is more than half of v_{IN}. But since d_D is d_E's opposite fraction, d_D is less than 50%, which means $v_{SW(AVG)}$ is also less than half of v_O. Interestingly, only a v_O that is higher than v_{IN} can establish a $v_{SWO(AVG)}$ that matches $v_{SWI(AVG)}$. The opposite is true when d_E is less than 50%: $v_{SW(AVG)}$ is less than half of v_{IN} and more than half of v_O, so v_O is lower than v_{IN}. In other words, L_X bucks when d_E is less than 50% (and d_D is greater than 50%) and boosts when d_E is greater than 50% (and d_D is less than 50%).

Example 5: Determine d_E when v_{IN} is 2 V and v_O is 1 V.

Solution:

$$d_E = \frac{v_D}{v_E + v_D} = \frac{v_O}{v_{IN} + v_O} = \frac{1}{2+1} = 33\%$$

Example 6: Determine d_E when v_{IN} is 2 V and v_O is 4 V.

Solution:

$$d_E = \frac{v_D}{v_E + v_D} = \frac{v_O}{v_{IN} + v_O} = \frac{4}{2+4} = 67\%$$

C. Power

Since v_{IN} connects to L_X a t_E fraction of t_C, v_{IN} sources a similar fraction of $i_{L(AVG)}$. v_{IN} therefore supplies an *input duty-cycle* d_{EI} fraction of $i_{L(AVG)}$:

$$P_{IN} = v_{IN} i_{IN(AVG)} = v_{IN} i_{L(AVG)} d_E = v_{IN}\left(\frac{i_{O(AVG)}}{d_D}\right) d_E = v_{IN}\left(\frac{i_{O(AVG)}}{d_{DO}}\right) d_{EI}. \quad (31)$$

The output similarly receives a d_D and *output duty-cycle* d_{DO} fraction of $i_{L(AVG)}$, which means $i_{L(AVG)}$ is $1/d_D$ times greater than $i_{O(AVG)}$ and P_O is

$$P_O = v_O i_{O(AVG)} = v_O i_{L(AVG)} d_D = v_O i_{L(AVG)} d_{DO}. \quad (32)$$

$i_{L(AVG)}$ is therefore greater than both $i_{IN(AVG)}$ and $i_{O(AVG)}$.

3.2. Asynchronous Buck–Boost

A. Power Stage

L_X in the asynchronous buck–boost energizes with transistors and drains with diodes. The input and ground energize switches S_{EI} and S_{EG} in Fig. 9 are therefore transistors M_{EI} and M_{EG} in the asynchronous buck–boost in Fig. 10. The connectivity of the switches and the gate drive that the circuit can establish ultimately dictate which type of MOSFET is more effective for M_{EI} and M_{EG}.

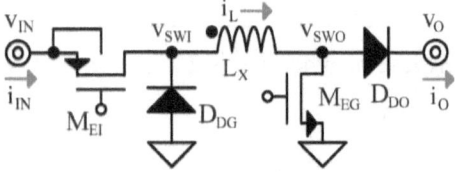

Fig. 10. Asynchronous buck–boost.

When closed, M_{EI}'s source and drain voltages v_S and v_D are close to v_{IN}. Since v_O is not always greater than v_{IN}, M_{EI}'s gate voltage v_G can swing reliably only between v_{IN} and ground. Under these conditions, the maximum gate drive of an NFET would be v_{GST}, $v_G - v_S - v_{TN}$, $v_{IN} - v_{IN} - v_{TN}$, or just $-v_{TN}$, which is too low to close an NFET. The maximum gate drive of a PFET would be v_{SGT}, $v_S - v_G - |v_{TP}|$, $v_{IN} - 0 - |v_{TP}|$, or $v_{IN} - |v_{TP}|$, which is greater and more likely to close a PFET. This is why M_{EI} is usually a PFET, because a high v_S calls for a high-side switch.

Since PMOS source terminals receive current, v_{IN} connects to M_{EI}'s source and M_{EI}'s source arrow points into the transistor (into the channel through which i_L flows). To ensure M_{EI}'s body diode does not energize L_X by conducting i_{IN} when v_{SWI} falls, the N-type bulk terminal should short the body diode that conducts i_{IN}. M_{EI}'s v_B should therefore connect to v_{IN}.

When closed, M_{EG}'s source and drain voltages v_S and v_D are close to ground. The only type of MOSFET that can close when v_S is the most negative potential like this is N channel because P-channel devices need a gate voltage that is lower than v_S to close. M_{EG} is therefore a low-side NFET with a source that connects and points to ground (away from the channel through which i_L flows) because N-channel source terminals output current.

The P-type bulk terminal connects to ground to ensure M_{EG}'s body diode does not energize L_X by conducting i_{IN}. Although Fig. 10 does not show this connection, the connection is there. NFETs in P-type substrates normally sit directly over the substrate, so the substrate *is* M_{EG}'s bulk terminal. To isolate all devices in the *integrated circuit* (IC), P-type substrates normally connect to ground, the most negative potential.

After M_{EI} and M_{EG} energize L_X, i_L flows to the right. The diodes that implement the ground and output drain switches S_{DG} and S_{DO} must therefore derive i_L from ground and direct it to v_O. In other words, the direction of i_L sets the orientation of D_{DG} and D_{DO} in Fig. 10.

B. Duty-Cycle Translation

Since D_{DG} drops a diode voltage below ground and D_{DO} a diode voltage above v_O, L_X drains with v_O plus two diode voltages. This higher v_D drains L_X faster than in the ideal buck–boost, so t_D and t_C are shorter and t_E is therefore a larger fraction of t_C. In other words, d_E in the asynchronous buck–boost is higher than in the ideal buck–boost.

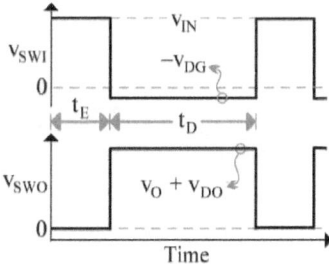

Fig. 11. Asynchronous buck–boost voltages.

v_{SWI} swings between v_{IN} and $-v_{DG}$ and v_{SWO} swings between zero and $v_O + v_{DO}$, like Fig. 11 illustrates. As a result, v_{SWI} connects to v_{IN} a t_E fraction of t_C or d_E' and to $-v_{DG}$ a t_D fraction d_D', so $v_{SWI(AVG)}$ is

$$v_{SWI(AVG)} = v_{IN}d_E' - v_{DG}d_D' = v_{IN}d_E' - v_{DG}\left(1 - d_E'\right). \qquad (33)$$

v_{SWO} connects to zero a t_E fraction d_E' and to $v_O + v_{DG}$ a t_D fraction d_D', so $v_{SWO(AVG)}$ is

$$v_{SWO(AVG)} = (0)d_E' + \left(v_O + v_{DO}\right)d_D' = \left(v_O + v_{DO}\right)\left(1 - d_E'\right). \qquad (34)$$

Since the averages match, d_E'/d_D' is higher than the ideal d_E/d_D by the fraction of v_{IN} that two diode voltages represents:

$$\frac{d_E'}{d_D'} = \frac{v_D}{v_E} = \frac{v_O + v_{DG} + v_{DO}}{v_{IN}} = \frac{v_O}{v_{IN}} + \frac{v_{DG} + v_{DO}}{v_{IN}}. \tag{35}$$

As a result, the diodes raise the buck-to-boost transition point of d_E'/d_D' (when v_O matches v_{IN}) from v_O/v_{IN}'s one to one plus two diode fractions of v_{IN}. Since d_D' is $1 - d_E'$, the diodes raise v_D's v_O in the numerator of d_E' by a larger fraction than in the denominator that v_E's v_{IN} and v_D's v_O set:

$$d_E' = \frac{v_D}{v_E + v_D} = \frac{v_O + v_{DO} + v_{DG}}{v_{IN} + v_O + v_{DG} + v_{DO}} > d_E. \tag{36}$$

So the effect of the diodes is to increase d_E to d_E'.

Example 7: Determine d_E' when v_{IN} is 2 V, v_O is 4 V, and D_{DG} and D_{DO} drop 800 mV.

Solution:

$$d_E' = \frac{v_O + v_{DO} + v_{DG}}{v_{IN} + v_O + v_{DG} + v_{DO}} = \frac{4 + 800m + 800m}{2 + 4 + 800m + 800m} = 74\%$$

Note: d_E' is higher than d_E in the ideal example because D_{DG} and D_{DO} increase L_X's drain voltage. So t_D shortens and t_E's fraction of t_C's $t_D + t_E$ increases.

C. Conduction Modes

In continuous conduction, i_L rises and falls about the $i_{L(AVG)}$ that keeps i_L above zero, like the thinner traces of i_L in Fig. 12 illustrate. As the controller adjusts how much current v_{IN} supplies or v_O sinks, $i_{L(AVG)}$ shifts. i_L reaches zero at the end of t_{SW} when $i_{L(AVG)}$ is half the ripple Δi_L and before t_{SW} ends when $i_{L(AVG)}$ is below that level. Since the diodes cannot conduct reverse current, i_L remains around zero until the next t_{SW}. L_X is in discontinuous conduction this way.

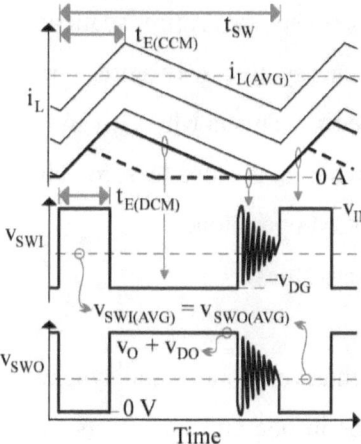

Fig. 12. Discontinuous-conduction waveforms.

Just before the diodes stop conducting (towards the end of t_D), v_{SWI} is a diode voltage below ground and v_{SWO} is a diode voltage above v_O. Parasitic capacitances at the switching nodes C_{SWI} and C_{SWO} (in Fig. 13) therefore hold energy when L_X depletes: $0.5C_{SWI}v_{DG}^2$ and $0.5C_{SWO}(v_O + v_{DO})^2$. So when all switches open (at the end of t_D) before the cycle t_{SW} elapses, v_{SWI} and v_{SWO} impress $v_O + v_{DG} + v_{DO}$ across L_X. This first drains C_{SWO} and C_{SWI} into L_X, and after v_{SWI} reaches zero, drains C_{SWO} into L_X and C_{SWI}. v_{SWO} therefore drops as v_{SWI} climbs. v_L then reverses when v_{SWO} crosses and falls below v_{SWI}. So past this point, C_{SWO} *and* L_X drain into C_{SWI}.

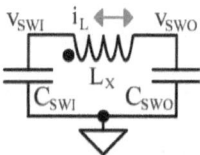

Fig. 13. Drained and disconnected asynchronous buck–boost inductor.

After L_X depletes, v_{SWI} is greater than v_{SWO}. L_X therefore draws a current from C_{SWI} that drains C_{SWI} and energizes L_X and C_{SWO}. L_X stops energizing when v_{SWO} and v_{SWI} cross. i_L nevertheless continues to flow to, this time, drain C_{SWI} *and* L_X into C_{SWO}. When L_X depletes, v_{SWO} is again greater than v_{SWI}, so the entire sequence repeats. L_X and the capacitors exchange energy this way until series resistances burn the energy or a new t_{SW} begins. This is why v_{SWI} and v_{SWO} in Fig. 12 oscillate about their averages when i_L is close to zero. These damped oscillations are characteristic of DCM operation.

3.3. Synchronous Buck–Boost

A. Power Stage

The difference between asynchronous and synchronous power supplies is what drains L_X: diodes in one and transistors in the other. So the FETs M_{EI} and M_{EG} that energize the asynchronous buck–boost and implement S_{EI} and S_{EG} in the ideal case also energize the synchronous counterpart in Fig. 14. The difference in Fig. 14 is that FETs M_{DG} and M_{DO} drain L_X, which is the function of S_{DG} and S_{DO} and D_{DG} and D_{DO} in the ideal and asynchronous cases. The connectivity of the switches and the gate drive v_{GST} that the circuit can establish ultimately dictate which type of MOSFET is more effective for M_{DG} and M_{DO}.

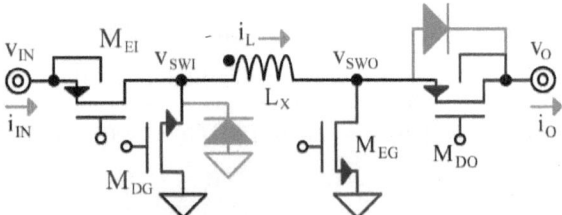

Fig. 14. Synchronous buck–boost.

Since digital gate signals can crisscross, input and output power FETs can momentarily short v_{IN} and v_O to ground. To avoid this, the controller inserts dead times t_{DT} between the conduction periods of adjacent switches: between M_{EI}'s and M_{DG}'s and between M_{EG}'s and M_{DO}'s. Body diodes must therefore connect in such a way that they not only block reverse inductor current but, if needed, also steer i_L across these t_{DT}'s.

When closed, M_{DG}'s source and drain voltages v_S and v_D are close to ground. The only type of MOSFET that can close when v_S is the most negative potential like this is N channel because P-channel devices need a gate voltage that is lower than v_S to close. M_{DG} is therefore a low-side NFET with a source that connects and points to v_{SWI} (away from the channel through which i_L flows) because N-channel source terminals output current.

The P-type bulk terminal connects to ground to ensure M_{DG}'s body diode does not conduct reverse i_O. Although Fig. 14 does not show this connection, the connection is there because NFETs in P-type substrates normally sit directly over a grounded substrate. Besides, this connection ensures the body diode can steer i_L across t_{DT}'s.

When closed, M_{DO}'s source and drain voltages v_S and v_D are close to v_O. Since v_{IN} is not always greater than v_O, M_{DO}'s gate voltage v_G can swing reliably only between v_O and ground. Under these conditions, the maximum gate drive of an NFET would be v_{GST}, $v_G - v_S - v_{TN}$, $v_O - v_O - v_{TN}$, or just $-v_{TN}$, which is too low to close an NFET. The maximum gate drive of a PFET would be v_{SGT}, $v_S - v_G - |v_{TP}|$, $v_O - 0 - |v_{TP}|$, or $v_O - |v_{TP}|$, which is greater and more likely to close the switch. This is why M_{DO} is oftentimes a PFET, because a high v_S calls for a high-side switch.

Since PMOS source terminals receive current, L_X feeds M_{DO}'s source and M_{DO}'s source arrow points into the transistor (into the channel through which i_L flows). To ensure M_{DO}'s body diode does not conduct reverse i_O when v_{SWO} falls, the N-type bulk terminal connects to v_O. This connection also allows the remaining body diode to steer dead-time currents into v_O.

B. Duty-Cycle Translation

The difference between asynchronous and synchronous operation across t_C is that M_{DG} and M_{DO} drop lower voltages than D_{DG} and D_{DO}. So v_{SWI} in Fig. 15 is close to zero and v_{SWO} close to v_O when M_{DG} and M_{DO} close across t_D. But since M_{DG} and M_{DO} open across the dead-time portions of t_D, D_{DG} drops v_{SWI} to $-v_{DG}$ and D_{DO} raises v_O to $v_O + v_{DO}$ across t_{DT}'s.

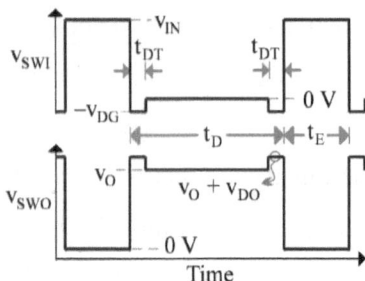

Fig. 15. Continuous conduction with non-reversing inductor current.

This means that L_X drains two t_{DT} fractions of t_C with v_O plus two diode voltages and the remainder of t_D's fraction with v_O. This higher v_D drains L_X faster than in the ideal buck–boost, but not as fast as in the asynchronous implementation (because v_D is higher only momentarily). t_D and t_C are therefore shorter, but not as short as in the asynchronous buck–boost. t_E is similarly a longer fraction of t_C, but also not as long a fraction as the asynchronous d_E is.

v_{SWI} connects to v_{IN} a t_E fraction of t_C, $-v_{DG}$ two t_{DT} fractions, and zero for the fraction of t_D that remains. So v_{SWI}'s average is

$$V_{SWI(AVG)} = V_{IN} d_E'' - V_{DG} \left(\frac{2t_{DT}}{t_C} \right) + (0) \left(\frac{t_D - 2t_{DT}}{t_C} \right). \qquad (37)$$

v_{SWO} connects to zero a t_E fraction of t_C, $v_O + v_{DG}$ two t_{DT} fractions, and v_O across what remains of t_D:

$$V_{SWO(AVG)} = (0) d_E'' + \left(v_O + v_{DO} \right) \left(\frac{2t_{DT}}{t_C} \right) + v_O \left(\frac{t_D - 2t_{DT}}{t_C} \right)$$

$$= v_o d_D'' + v_{DO}\left(\frac{2t_{DT}}{t_C}\right) = v_o\left(1 - d_E''\right) + v_{DO}\left(\frac{2t_{DT}}{t_C}\right). \quad (38)$$

Since the averages match in steady state and d_D'' is $1 - d_E''$, the two dead-time fractions raise d_E by two diode fractions of $v_{IN} + v_O$:

$$d_E'' = \frac{v_O}{v_{IN} + v_O} + \left(\frac{v_{DO} + v_{DG}}{v_{IN} + v_O}\right)\left(\frac{2t_{DT}}{t_C}\right) = d_E + \left(\frac{v_{DO} + v_{DG}}{v_{IN} + v_O}\right)\left(\frac{2t_{DT}}{t_C}\right) > d_E. \quad (39)$$

Example 8: Determine d_E'' when v_{IN} is 2 V, v_O is 4 V, MOS diodes drop 800 mV, t_{DT} is 50 ns, and t_C is 1 μs.

Solution:

$$d_E'' = \frac{v_O}{v_{IN} + v_O} + \left(\frac{v_{DO} + v_{DG}}{v_{IN} + v_O}\right)\left(\frac{2t_{DT}}{t_C}\right)$$

$$= \frac{4}{2+4} + \left(\frac{800m + 800m}{2+4}\right)\left[\frac{2(50n)}{1\mu}\right] = 69\%$$

Note: d_E'' is higher than d_E in the ideal example because D_{DG} and D_{DO} raise v_D. But since the diodes do so only two t_{DT} fractions of t_C, t_D shortens and t_E's fraction of t_C climbs less than d_E' in the asynchronous example.

C. Conduction Modes

The core difference between asynchronous and synchronous operation hinges on the controller. If the controller opens and closes the synchronous drain transistors M_{DG} and M_{DO} when the asynchronous diodes D_{DG} and D_{DO} naturally would, the only difference is the voltage dropped across the switches: millivolts with FETs and 0.60–0.75 V with diodes. But since M_{DG} and M_{DO} are off across dead-time periods, M_{DG}'s and M_{DO}'s body diodes in synchronous converters conduct like their asynchronous counterparts across t_{DT}'s.

If the controller does not open the drain FETs when i_L reaches zero, discontinuous conduction is not possible with the synchronous converter. The reason for this is FETs are bidirectional. So if i_L reaches zero before t_{SW} ends, the drain voltage that M_{DG} and M_{DO} impresses across L_X induces i_L to fall below zero like the thicker trace in Fig. 16 shows. In other words, i_L reverses and pulls current from v_O (in Fig. 14).

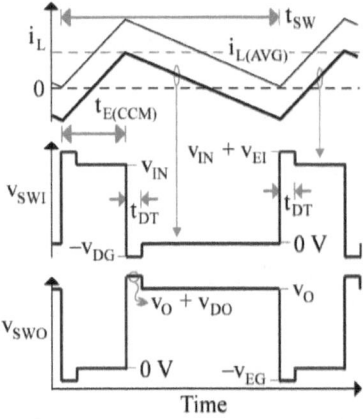

Fig. 16. Continuous conduction with reversing inductor current.

Since M_{DG}'s and M_{DO}'s body diodes are not bidirectional, reverse dead-time i_L does not flow through these diodes. Instead, negative i_L flows through M_{EG}'s and M_{EI}'s body diodes. So when M_{DG} and M_{DO} open before starting another t_{SW}, v_{SWO} falls to $-v_{EG}$ and v_{SWI} rises to $v_{IN} + v_{EI}$ across t_{DT}, like Fig. 16 shows, instead of rising to $v_O + v_{DO}$ and falling to $-v_{DG}$ like a positive i_L induces.

The worst part about this is that L_X pulls power from v_O and returns it to v_{IN} when i_L reverses, which is the opposite of what a power supply should do. This is why designers often open M_{DG} and M_{DO} when i_L reaches zero, like diodes would. By forcing discontinuous conduction this way, the system does not burn Ohmic power to deliver and return energy that v_O does not receive.

Note one t_{DT} is in t_D and another in t_E when i_L reverses, whereas without negative conduction, t_D includes both t_{DT}'s. And since v_{SWI} and v_{SWO} rise and fall by a similar diode voltage across similar t_{DT}'s, diode effects on $v_{SWI(AVG)}$ and $v_{SWO(AVG)}$ tend to cancel. So with negative conduction, d_E is close to the ideal case (lower than d_E'').

D. Diode Conduction

M_{DG}'s and M_{DO}'s body diodes conduct dead-time i_L like D_{DG} and D_{DO} in Fig. 10. In other words, the synchronous network operates like the asynchronous converter across t_{DT}'s. If MOSFET threshold voltages are lower than diode voltages, v_{SWI} falls below M_{DG}'s grounded drain and gate terminals until M_{DG}'s implicit diode action drops v_{GS} and conducts. v_{SWO} similarly climbs above M_{DO}'s v_O-supplied gate and drain terminals until M_{DO}'s resulting diode connection drops v_{SG} and conducts. So M_{DG}'s and M_{DO}'s body diodes do not conduct into the substrate or M_{DO}'s body, which would otherwise inject noise power into the shared substrate.

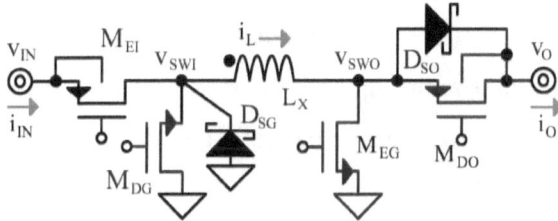

Fig. 17. Schottky-clamped synchronous buck–boost.

When thresholds are higher than diode voltages and noise power is unacceptably high, designers connect Schottky diodes across M_{DG}'s and M_{DO}'s body diodes like Fig. 17 shows. This way, with the lower voltages that Schottkys require to conduct, the Schottkys conduct most of the dead-time current. Steering current away from the body diodes this way reduces the noise-producing current that would otherwise flow through the substrate.

4. Buck

4.1. Ideal Buck

A. Power Stage

As the name implies, buck SLs *buck* v_{IN} to a lower v_O. Since v_{IN} is always greater than v_O, $v_{IN} - v_O$ can be the v_E that energizes L_X. v_{IN} energizes L_X into v_O directly this way. This is good because v_{IN} delivers energy with i_L into v_O that L_X does not need to carry and transfer. As a result, i_L peaks to a lower value than in the buck–boost. And with a lower i_L, series resistances burn less power, so P_O in a buck is usually a higher fraction of P_{IN} than P_O is in a buck–boost.

The switches that connect L_X to v_O in the buck–boost in Fig. 9 disappear in the buck of Fig. 18 because L_X both energizes and drains to v_O. Here, energize switch S_{EI} draws input power by connecting L_X across v_{IN} and v_O. S_{EI} then opens and ground drain switch S_{DG} closes to connect L_X's input terminal to ground. This way, v_L's polarity reverses to $-v_O$ to drain L_X into v_O. v_L (in Fig. 5) therefore swings between $+v_E$'s $v_{IN} - v_O$ and $-v_D$'s $-v_O$ and i_L ramps up with $+v_E$'s $v_{IN} - v_O$ and down with $-v_D$'s $-v_O$. Not surprisingly, the buck is basically the part of the buck–boost that bucks: S_{EI}, S_{DG}, and L_X.

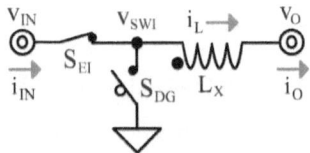

Fig. 18. Ideal buck.

B. Duty-Cycle Translation

In steady state, v_L's average is zero, which means the average of the switching voltage v_{SWI} matches v_O. Since v_{IN} connects to v_{SWI} a t_E

fraction of t_C, v_{SWI}'s average and matching v_O are a duty-cycled fraction of v_{IN}:

$$v_{SWI(AVG)} = v_{IN}\left(\frac{t_E}{t_{SW}}\right) + (0)\left(\frac{t_D}{t_{SW}}\right) = v_{IN}d_E = v_O.$$ (40)

d_E is therefore a v_O fraction of v_{IN}:

$$d_E = \frac{v_D}{v_E + v_D} = \frac{v_O}{v_{IN}},$$ (41)

like $v_{IN} - v_O$ for v_E and v_O for v_D in the general expression predict.

Example 9: Determine d_E when v_{IN} is 2 V and v_O is 1 V.
Solution:

$$d_E = \frac{v_D}{v_E + v_D} = \frac{v_O}{v_{IN}} = \frac{1}{2} = 50\%$$

C. Power

Since v_{IN} connects to L_X a t_E fraction of t_C, v_{IN} sources a similar fraction of $i_{L(AVG)}$. v_{IN} therefore supplies a d_{EI} fraction of $i_{L(AVG)}$:

$$P_{IN} = v_{IN}i_{IN(AVG)} = v_{IN}i_{L(AVG)}d_E = v_{IN}i_{O(AVG)}d_{EI}.$$ (42)

The output receives i_L continuously across t_C, so v_O receives i_L's average, which means $i_{O(AVG)}$ is $i_{L(AVG)}$, d_{DO} is one, and P_O is

$$P_O = v_O i_{O(AVG)} = v_O i_{L(AVG)}d_{DO} = v_O i_{L(AVG)}.$$ (43)

4.2. Asynchronous Buck

A. Power Stage

L_X in the asynchronous buck energizes with a transistor and drains with a diode. The energize switch S_{EI} in Fig. 18 is therefore a transistor M_{EI} in Fig. 19 and the drain switch S_{DG} is a diode D_{DG}. Since i_L flows to the right after M_{EI} energizes L_X, D_{DG} must derive i_L from ground and direct it to v_O. This is why D_{DG}'s anode connects to ground and D_{DG}'s cathode to

v_{SWI}. This circuit is basically the part of the asynchronous buck–boost in Fig. 10 that bucks: M_{EI}, D_{DG}, and L_X. So the selection and connectivity of M_{EI} here matches that of M_{EI} in the buck–boost.

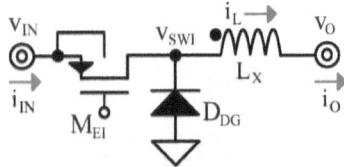

Fig. 19. Asynchronous buck.

B. Duty-Cycle Translation

Since D_{DG} drops a diode voltage below ground, L_X drains with v_O plus a diode voltage. This higher v_D drains L_X faster than in the ideal buck, so t_D and t_C are shorter and t_E is a larger fraction of t_C. In other words, d_E in the asynchronous buck is higher than in the ideal buck.

v_{SWI} swings between v_{IN} and $-v_{DG}$, like in Figs. 10 and 11. v_{SWI} therefore connects to v_{IN} a t_E fraction of t_C and to $-v_{DG}$ a t_D fraction $d_D{}'$, so v_{SWI}'s average and matching v_O are

$$v_{SWI(AVG)} = v_{IN}d_E{}' - v_{DG}d_D{}' = v_{IN}d_E{}' - v_{DG}\left(1 - d_E{}'\right) = v_O. \tag{44}$$

Since $d_D{}'$ is $1 - d_E{}'$ and v_O is less than v_{IN} in a buck, the diode raises v_D's v_O in the numerator of $d_E{}'$ by a larger fraction than in the denominator that v_E's $v_{IN} - v_O$ and v_D's v_O set:

$$d_E{}' = \frac{v_D}{v_E + v_D} = \frac{v_O + v_{DG}}{v_{IN} + v_{DG}} > d_E. \tag{45}$$

So the effect of the diode is to increase d_E.

Example 10: Determine $d_E{}'$ when v_{IN} is 2 V, v_O is 1 V, and D_{DG} drops 800 mV.

Solution:

$$d_E' = \frac{v_O + v_{DG}}{v_{IN} + v_{DG}} = \frac{1 + 800m}{2 + 800m} = 64\%$$

Note: d_E' is higher than d_E in the ideal example because D_{DG} increases L_X's drain voltage. So t_D shortens and t_E's fraction of t_C's $t_D + t_E$ increases.

C. Conduction Modes

In static applications, the output is either a battery or a load that requires a steady v_O or i_O. Designers normally add capacitance to the output of voltage regulators and LED drivers because controllers cannot respond instantaneously. So either way, C_O is intentional and therefore much higher than the parasitic capacitance C_{SWI} present at the switching node.

In continuous conduction, i_L climbs and falls about the $i_{L(AVG)}$ that keeps i_L above zero. As the controller adjusts how much current v_{IN} supplies or v_O sinks, $i_{L(AVG)}$ shifts. Like in the asynchronous buck–boost of Fig. 12, i_L reaches zero at the end of t_{SW} when $i_{L(AVG)}$ is half the ripple Δi_L and before t_{SW} ends when $i_{L(AVG)}$ is below that level. Since D_{DG} cannot conduct reverse current, i_L remains zero until the next t_{SW}. This is discontinuous conduction.

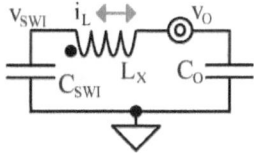

Fig. 20. Drained and disconnected asynchronous buck inductor.

Just before the diode stops conducting (towards the end of t_D), v_{SWI} is a diode voltage below ground. C_{SWI} in Fig. 20 therefore holds energy $0.5C_{SWI}v_{DG}^2$ when L_X depletes. So when the switches open (at the end of t_D) before the cycle elapses, v_O and v_{SWI} impress $v_O + v_{DG}$ across L_X. This drains C_{SWI} and C_O into L_X. After v_{SWI} rises to ground, C_O drains into both L_X and C_{SWI}. L_X stops energizing when v_{SWI} reaches v_O.

Since L_X holds energy when v_{SWI} reaches v_O, i_L continues to charge C_{SWI} until L_X depletes. C_{SWI} therefore charges above v_O when L_X depletes. The voltage that v_{SWI} and v_O now impress across L_X draws a reverse current that first drains C_{SWI} into L_X and C_O and then (when v_{SWI} falls below v_O) drains both C_{SWI} and L_X into C_O. When L_X depletes, v_O is again greater than v_{SWI} and the entire sequence repeats. L_X and the capacitors exchange energy this way until series resistances burn the energy or a new t_{sw} begins. This is why v_{SWI} oscillates about its average v_O when i_L is zero like in the asynchronous buck–boost from Fig. 12.

4.3. Synchronous Buck

A. Power Stage

The difference between asynchronous and synchronous power supplies is what drains L_X: diodes in one and transistors in the other. So the FET M_{EI} that energizes the asynchronous buck and implements S_{EI} in the ideal buck also energizes the synchronous counterpart in Fig. 21. The difference in Fig. 21 is that transistor M_{DG} drains L_X, which is the function of S_{DG} and D_{DG} in the ideal and asynchronous bucks. This circuit is the part of the synchronous buck–boost in Fig. 14 that bucks: M_{EI}, M_{DG}, and L_X. So the selection and connectivity of M_{DG} and behavior of v_{SWI} in continuous and discontinuous conduction match those of M_{DG} and v_{SWI} in the synchronous buck–boost.

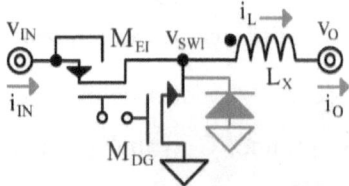

Fig. 21. Synchronous buck.

B. Duty-Cycle Translation

The difference between asynchronous and synchronous operation across t_C is that M_{DG} drops a lower voltage than D_{DG}. So v_{SWI} is close to zero when M_{DG} closes across t_D. But since M_{DG} and M_{DO} open across the dead-time portions of t_D, v_{SWI} reaches $-v_{DG}$ across t_{DT}'s, like v_{SWI} in the synchronous buck–boost from Fig. 15.

This means that L_X drains two t_{DT} fractions of t_C with v_O plus one diode voltage and the remainder of t_D's fraction with v_O. This higher v_D drains L_X faster than in the ideal buck, but not as fast as in the asynchronous implementation because v_D is higher only momentarily. t_D and t_C are therefore shorter, but not as short as in the asynchronous buck. t_E is similarly a longer fraction of t_C, but also not as long a fraction as the asynchronous d_E' is.

v_{SWI} connects to v_{IN} a t_E fraction of t_C, $-v_{DG}$ two t_{DT} fractions, and zero for the fraction of t_D that remains. So v_{SWI}'s average and matching v_O are

$$V_{SWI(AVG)} = V_{IN}d_E'' - V_{DG}\left(\frac{2t_{DT}}{t_C}\right) + (0)\left(\frac{t_D - 2t_{DT}}{t_C}\right) = v_O. \qquad (46)$$

Since v_O matches v_{SWI}'s average in steady state, the two dead-time fractions raise d_E by a diode fraction of v_{IN}:

$$d_E'' = \frac{V_O}{V_{IN}} + \frac{V_{DG}}{V_{IN}}\left(\frac{2t_{DT}}{t_C}\right) = d_E + \frac{V_{DG}}{V_{IN}}\left(\frac{2t_{DT}}{t_C}\right) > d_E. \qquad (47)$$

Example 11: Determine d_E'' when v_{IN} is 2 V, v_O is 1 V, D_{DG} drops 800 mV, t_{DT} is 50 ns, and t_C is 1 μs.

Solution:

$$d_E'' = \frac{v_O}{v_{IN}} + \left(\frac{v_{DG}}{v_{IN}}\right)\left(\frac{2t_{DT}}{t_C}\right) = \frac{1}{2} + \left(\frac{800m}{2}\right)\left[\frac{2(50n)}{1\mu}\right] = 54\%$$

Note: d_E'' here is higher than d_E in the ideal example because D_{DG} raises v_D. But since it does so only two t_{DT} fractions of t_C, t_D and t_C shorten and t_E's fraction of t_C rises less than d_E' in the asynchronous example.

C. Conduction Modes

Since the buck is the part of the buck–boost that bucks, M_{EI} and M_{DG} switch the same way and produce the same v_{SWI}. So like in the buck–boost, the synchronous buck operates like the asynchronous counterpart when the controller opens and closes M_{DG} like D_{DG} naturally would. Except, M_{DG} drops millivolts and D_{DG} drops 0.60–0.75 V. But since M_{DG} is off across t_{DT}'s, M_{DG}'s body diode emulates D_{DG} across t_{DT}'s.

If the controller does not open M_{DG} when i_L falls to zero before t_{SW} lapses, M_{DG} lets i_L reverse direction. Since M_{DG}'s body diode cannot conduct reverse current, M_{EI}'s body diode steers this negative i_L across the t_{DT} that follows to v_{IN}. So v_{SWI} climbs to $v_{IN} + v_{EI}$ like Fig. 16 shows. Returning energy to v_{IN} this way means L_X transfers and burns more power than necessary.

Note one t_{DT} is in t_D and another in t_E when i_L reverses, whereas without negative conduction, t_D includes both t_{DT}'s. And since v_{SWI} rises and falls by a similar diode voltage across similar t_{DT}'s, diode effects on $v_{SWI(AVG)}$ tend to cancel. So with negative conduction, d_E is close to the ideal case (lower than d_E'').

D. Diode Conduction

If M_{DG}'s threshold voltage is lower than a diode voltage, M_{DG} closes across t_{DT}'s when v_{SWI} falls a v_{TN} below M_{DG}'s grounded gate voltage. M_{DG}'s body diode does not steer current into the substrate when this happens. A Schottky diode across M_{DG} similarly steals dead-time current away from the body diode and the substrate into which the body diode conducts.

5. Boost

5.1. Ideal Boost

A. Power Stage

As the name implies, boost SLs *boost* v_{IN} to a higher v_O. Since v_O is always greater than v_{IN}, $v_O - v_{IN}$ can be the v_D that drains L_X. This way, v_{IN} supplies power as L_X drains. In other words, v_{IN} delivers energy with i_L into v_O that L_X does not need to carry. i_L therefore peaks to a lower value than in the buck–boost, which means series resistances burn less power and P_O in a boost is, as a result, usually a higher fraction of P_{IN} than P_O is in a buck–boost.

The switches that connect v_{IN} to L_X in the buck–boost in Fig. 9 disappear in the boost of Fig. 22 because L_X both energizes and drains from v_{IN}. So ground energize switch S_{EG} draws input power by connecting L_X across v_{IN} and ground. When done energizing, S_{EG} opens and output drain switch S_{DO} closes to connect L_X's output terminal to v_O. This way, v_L's polarity is v_{IN} when energizing L_X and inverts to $v_{IN} - v_O$ when draining L_X into v_O. v_L (in Fig. 5) therefore swings between $+v_E$'s v_{IN} and $-v_D$'s $v_{IN} - v_O$ and i_L ramps up with $+v_E$'s v_{IN} and down with $-v_D$'s $v_{IN} - v_O$. Not surprisingly, the boost is basically the part of the buck–boost that boosts: L_X, S_{EG}, and S_{DO}.

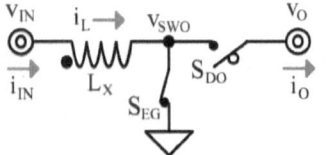

Fig. 22. Ideal boost.

B. Duty-Cycle Translation

In steady state, v_L's average is zero, which means the switching voltage v_{SWO} is, on average, equal to v_{IN}. Since v_O connects to v_{SWO} a t_D fraction of t_C, v_{IN} and v_{IN}'s matching $v_{SWO(AVG)}$ are a duty-cycled fraction of v_O:

$$V_{IN} = V_{SWO(AVG)} = (0)\left(\frac{t_E}{t_{SW}}\right) + v_O\left(\frac{t_D}{t_{SW}}\right) = v_O d_D = v_O\left(1 - d_E\right), \qquad (48)$$

which is why v_O is greater than v_{IN} by the multiplying scalar that $1/d_D$ sets:

$$V_O = \frac{V_{IN}}{d_D} = \frac{V_{IN}}{1 - d_E}. \qquad (49)$$

And since d_D is $1 - d_E$, d_E is a $v_O - v_{IN}$ fraction of v_O:

$$d_E = \frac{v_D}{v_E + v_D} = \frac{v_O - v_{IN}}{v_O} = 1 - \frac{v_{IN}}{v_O}, \qquad (50)$$

like v_{IN} for v_E and $v_O - v_{IN}$ for v_D in the general expression predict.

Example 12: Determine d_E when v_{IN} is 2 V and v_O is 4 V.

Solution:

$$d_E = \frac{v_D}{v_E + v_D} = \frac{v_O - v_{IN}}{v_O} = \frac{4 - 2}{4} = 50\%$$

C. Power

The input sources i_L continuously across t_C, so v_{IN} supplies i_L's average, d_{EI} is one, and P_{IN} is

$$P_{IN} = V_{IN}i_{IN(AVG)} = V_{IN}i_{L(AVG)}d_{EI} = V_{IN}\left(\frac{i_{O(AVG)}}{d_D}\right) = V_{IN}\left(\frac{i_{O(AVG)}}{d_{DO}}\right). \qquad (51)$$

Since v_O connects to L_X a t_D fraction of t_C, v_O sinks a d_D and d_{DO} fraction of $i_{L(AVG)}$, which means $i_{L(AVG)}$ is $1/d_D$ times greater than $i_{O(AVG)}$. v_O therefore receives the same P_O that the buck–boost converter receives: $v_Oi_{O(AVG)}$, $v_Oi_{L(AVG)}d_D$, and $v_Oi_{L(AVG)}d_{DO}$.

5.2. Asynchronous Boost

A. Power Stage

L_X in the asynchronous boost energizes with a transistor and drains with a diode. The ground energize switch S_{EG} in Fig. 22 is therefore a transistor M_{EG} in Fig. 23 and the output drain switch S_{DO} is a diode D_{DO}. Since i_L flows to the right after M_{EG} energizes L_X, D_{DO} steers i_L into v_O. This circuit is basically the part of the asynchronous buck–boost in Fig. 10 that boosts: L_X, M_{EG}, and D_{DO}. So the selection and connectivity of M_{EG} matches that of M_{EG} in the asynchronous buck–boost.

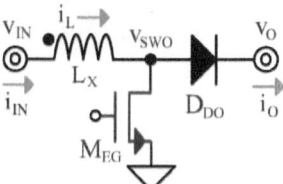

Fig. 23. Asynchronous boost.

B. Duty-Cycle Translation

Since D_{DO} drops a diode voltage above v_O, L_X drains with $v_O - v_{IN}$ plus a diode voltage. This higher v_D drains L_X faster than in the ideal boost, so t_D and t_C are shorter and t_E is a larger fraction of t_C. In other words, d_E is higher than in the ideal boost.

As in the buck–boost, v_{SWO} swings between ground and $v_O + v_{DO}$, like Fig. 11 shows. v_{SWO} therefore connects to ground a t_E fraction of t_C and to $v_O + v_{DO}$ a t_D fraction d_D', so v_{IN} and v_{IN}'s matching $v_{SWO(AVG)}$ are

$$V_{IN} = V_{SWO(AVG)} = (0)d_E' + \left(v_O + v_{DO}\right)d_D' = \left(v_O + v_{DO}\right)\left(1 - d_E'\right). \qquad (52)$$

Since d_D' is $1 - d_E'$ and $v_O - v_{IN}$ is less than v_O, the diode raises v_D's $v_O - v_{IN}$ in the numerator of d_E' by a larger fraction than in the denominator that v_E's v_{IN} and v_D's $v_O - v_{IN}$ establish:

$$d_E' = \frac{V_D}{V_E + V_D} = \frac{V_O + V_{DG} - V_{IN}}{V_O + V_{DG}} > d_E. \qquad (53)$$

So the effect of the diodes is to increase d_E.

Example 13: Determine d_E when v_{IN} is 2 V, v_O is 4 V, and D_{DO} drops 800 mV.

Solution:

$$d_E' = \frac{V_O + V_{DG} - V_{IN}}{V_O + V_{DG}} = \frac{4 + 800m - 2}{4 + 800m} = 58\%$$

Note: d_E' here is higher than d_E in the ideal example because D_{DO} increases L_X's drain voltage. So t_D shortens and t_E's fraction of t_C's $t_D + t_E$ increases.

C. Conduction Modes

In static applications, the input is either a dc source or the output of a voltage regulator that feeds a steady v_{IN}. Designers normally add capacitance to the output of voltage regulators because controllers cannot respond instantaneously. So either way, v_{IN}'s equivalent C_{IN} is intentional and therefore much higher than the parasitic capacitance C_{SWO} present at the switching node.

In continuous conduction, i_L climbs and falls about the $i_{L(AVG)}$ that keeps i_L above zero. As the controller adjusts how much current v_{IN} supplies or v_O sinks, $i_{L(AVG)}$ shifts. Like in the asynchronous buck–boost from Fig. 12, i_L reaches zero at the end of t_{SW} when $i_{L(AVG)}$ is half the ripple Δi_L and before t_{SW} ends when $i_{L(AVG)}$ is below that level. Since D_{DO} cannot conduct reverse current, i_L remains zero until the next t_{SW}. L_X is in discontinuous conduction this way.

Just before the diode stops conducting (towards the end of t_D), v_{SWO} is a diode voltage above v_O. C_{SWO} (in Fig. 24) therefore holds energy $0.5C_{SWO}(v_O + v_{DO})^2$ when L_X depletes. So when the switches open (at the end of t_D) before the cycle elapses, v_{IN} and v_{SWO} impress $v_O + v_{DG} - v_{IN}$ across L_X, which drains C_{SWO} into L_X and C_{IN}. L_X stops energizing when v_{SWO} falls to v_{IN} (when v_L is zero). Past that point, i_L drains L_X and C_{SWO} into C_{IN}.

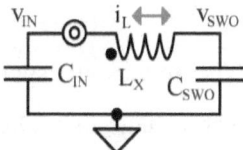

Fig. 24. Drained and disconnected asynchronous boost inductor.

Since v_{IN} is greater than v_{SWO} when L_X depletes, L_X draws a reverse current that first drains C_{IN} into L_X and C_{SWO} and then (when v_{SWO} rises above v_{IN}) drains C_{IN} *and* L_X into C_{SWO}. v_{SWO} is again greater than v_{IN} when L_X depletes, so the entire sequence repeats. L_X and the capacitors exchange energy this way until series resistances burn the energy or a new t_{SW} begins. This is why v_{SWO} oscillates when i_L is zero like v_{SWO} in the asynchronous buck–boost from Fig. 12.

5.3. Synchronous Boost

A. Power Stage

The difference between asynchronous and synchronous power supplies is what drains L_X: diodes in one and transistors in the other. So the ground FET M_{EG} that energizes the asynchronous boost and implements S_{EG} in the ideal case in Fig. 14 also energizes the synchronous counterpart in Fig. 25. The difference in Fig. 25 is that output transistor M_{DO} drains L_X, which is the function of S_{DO} and D_{DO} in the ideal and asynchronous boosts. This circuit is basically the part of the synchronous buck–boost that boosts: L_X, M_{EG}, and M_{DO}. So the selection and connectivity of M_{DO} and behavior of v_{SWO} in continuous and discontinuous conduction match those of M_{DO} and v_{SWO} in the synchronous buck–boost.

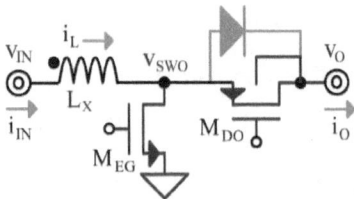

Fig. 25. Synchronous boost.

B. Duty-Cycle Translation

The difference between asynchronous and synchronous operation across t_D is that M_{DO} drops a lower voltage than D_{DO}. So v_{SWO} is close to v_O when M_{DO} closes within t_D. But since M_{DG} and M_{DO} open across the dead-time portions of t_D, v_{SWO} is $v_O + v_{DG}$ across the t_{DT}'s that t_D incorporates, like v_{SWO} in the synchronous buck–boost from Fig. 15.

This means that L_X drains two t_{DT} fractions of t_C with v_O plus one diode voltage and the remainder of t_D's fraction with v_O. This higher v_D drains L_X faster than in the ideal boost, but not as fast as in the asynchronous implementation because v_D is higher only momentarily. t_D and t_C are therefore shorter, but not as short as in the asynchronous boost. t_E is similarly a longer fraction of t_C, but also not as long a fraction as the asynchronous d_E' is.

v_{SWO} connects to zero a t_E fraction of t_C, $v_O + v_{DG}$ two t_{DT} fractions, and v_O for the fraction of t_D that remains. So $v_{SWO(AVG)}$ is

$$v_{IN} = v_{SWO(AVG)} = (0)d_E'' + \left(v_O + v_{DG}\right)\left(\frac{2t_{DT}}{t_C}\right) + v_O\left(\frac{t_D - 2t_{DT}}{t_C}\right). \quad (54)$$

Since v_{IN} matches v_{SWO}'s average in steady state and t_D/t_C is d_D'' or $1 - d_E''$, the two dead-time fractions raise d_E by a diode fraction of v_O:

$$d_E'' = \frac{v_O - v_{IN}}{v_O} + \left(\frac{V_{DO}}{v_O}\right)\left(\frac{2t_{DT}}{t_C}\right) = d_E + \left(\frac{V_{DO}}{v_O}\right)\left(\frac{2t_{DT}}{t_C}\right) > d_E. \qquad (55)$$

Example 14: Determine d_E'' when v_{IN} is 2 V, v_O is 4 V, D_{DO} drops 800 mV, t_{DT} is 50 ns, and t_C is 1 μs.

Solution:

$$d_E'' = \frac{v_O - v_{IN}}{v_O} + \left(\frac{V_{DO}}{v_O}\right)\left(\frac{2t_{DT}}{t_C}\right) = \frac{1}{2} + \frac{(0.8)(2)(50n)}{(4)(1\mu)} = 52\%$$

Note: d_E'' here is higher than d_E in the ideal example because D_{DO} raises v_D. But since the diode does so only two t_{DT} fractions of t_C, t_D shortens and t_E's fraction of t_C rises less than d_E' in the asynchronous example.

C. Conduction Modes

Since the boost is the part of the buck–boost that boosts, M_{EG} and M_{DO} switch the same way and produce the same v_{SWO}. So like in the buck–boost, the synchronous boost operates like the asynchronous counterpart when the controller opens and closes M_{DO} like D_{DO} naturally would. Except, M_{DO} drops millivolts and D_{DO} drops 0.60–0.75 V. But since M_{DO} is off across t_{DT}'s, M_{DO}'s body diode emulates D_{DO} across t_{DT}'s.

If the controller does not open M_{DO} when i_L falls to zero before t_{SW} lapses, M_{DO} lets i_L reverse direction. Since M_{DO}'s body diode cannot conduct reverse current, M_{EG}'s body diode steers this negative i_L to v_{IN} across the t_{DT} that follows. v_{SWO} therefore falls to $-v_{EG}$ across t_{DT} like Fig.

16 shows. Returning energy to v_{IN} this way means L_X transfers and burns more power than necessary.

Note one t_{DT} is in t_D and another in t_E when i_L reverses, whereas without negative conduction, t_D includes both t_{DT}'s. And since v_{SWO} rises and falls by a similar diode voltage across similar t_{DT}'s, diode effects on $v_{SWI(AVG)}$ tend to cancel. So with negative conduction, d_E is close to the ideal case (lower than $d_E"$).

D. Diode Conduction

If M_{DO}'s threshold voltage is lower than a diode voltage, M_{DO} closes across t_{DT}'s when v_{SWO} climbs a $|v_{TP}|$ above M_{DO}'s v_O-supplied gate voltage. M_{DO}'s body diode does not inject substrate current when this happens. A Schottky diode across M_{DO} similarly steals dead-time current away from the body diode and the substrate into which the parasitic *bipolar junction transistors* (BJTs) present inject current.

6. Flyback

6.1. Ideal Flyback

A. Power Stage

The flyback is an interesting variation of the buck–boost. Like all SLs, v_{IN} magnetizes the core of an inductor L_I. v_O similarly demagnetizes the core, but with another inductor L_O. In other words, L_I draws power from v_{IN} that a coupled L_O delivers to v_O.

The advantage of this setup is separate grounds for v_{IN} and v_O for what is termed *galvanic isolation*. This way, without a direct connection, stray noise currents do not couple. Ground levels can also be at different potentials. Galvanic isolation is ultimately a form of protection.

Aside from separate inductors, flybacks switch and operate like buck–boosts. L_I in Fig. 26, for example, magnetizes the core when the input switch S_{EI} connects v_{IN} across L_I. L_O demagnetizes the core after

S_{EI} opens when the output drain switch S_{DO} connects v_O across L_O. v_O drains the core because L_I and L_O couple in opposite directions, so v_{LO} is $-v_O$.

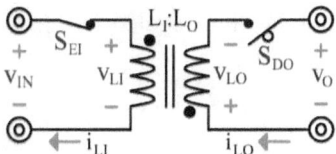

Fig. 26. Ideal (supply-switched) flyback.

L_I's i_{LI} in Fig. 27 therefore ramps up with the v_E that v_{LI}'s v_{IN} impresses across L_I and L_O's i_{LO} ramps down with the $-v_D$ that v_{LO}'s $-v_O$ impresses across L_O. Since S_{DO} opens across t_E and S_{EI} opens across t_D, i_{LO} is zero across t_E and i_{LI} is zero across t_D. As a result, v_{IN} couples a transformer translation of v_{IN} to L_O across t_E and v_O couples "back" a transformer translation of $-v_O$ to L_I across t_D. So when S_{EI} opens, v_{LI} practically "flies" from v_{IN} to $-v_O/k_L$ and v_{LO} from $v_{IN}k_L$ to $-v_O$. This "flyback" action on L_I is how this converter derives its name.

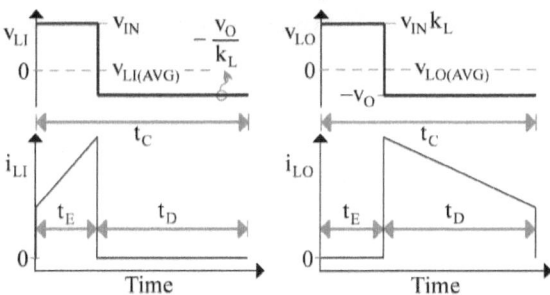

Fig. 27. Continuous-conduction waveforms in the flyback.

As a whole, the coupled inductors L_I:L_O operate and behave like L_X in the buck–boost. They energize and drain with v_{IN} and v_O. And the combined current they produce i_L or $i_{LI} + i_{LO}$ ripples about an average $i_{L(AVG)}$ that the controller adjusts like Fig. 5 shows.

B. Duty-Cycle Translation

In steady state, the average voltages across L_I and L_O are zero. Since v_{IN} is across L_I a t_E fraction of t_C and $-v_O/k_L$ couples back across L_I a t_D fraction, $v_{LI(AVG)}$ incorporates duty-cycled fractions of v_{IN} and $-v_O/k_L$:

$$v_{LI(AVG)} = v_{IN}\left(\frac{t_E}{t_C}\right) - \left(\frac{v_O}{k_L}\right)\left(\frac{t_D}{t_C}\right) = v_{IN}d_E - v_O\left(\sqrt{\frac{L_I}{L_O}}\right)d_D = 0. \quad (56)$$

Similarly, $v_{LO(AVG)}$ incorporates duty-cycled fractions of $v_{IN}k_L$ and $-v_O$ because $v_{IN}k_L$ couples across L_O a t_E fraction of t_C and $-v_O$ is across L_O a t_D fraction:

$$v_{LO(AVG)} = v_{IN}k_L\left(\frac{t_E}{t_C}\right) - v_O\left(\frac{t_D}{t_C}\right) = v_{IN}\left(\sqrt{\frac{L_O}{L_I}}\right)d_E - v_O d_D = 0. \quad (57)$$

Since d_D is $1 - d_E$, d_E is a v_O fraction of $v_{IN}k_L + v_O$:

$$d_E = \frac{v_D}{v_E + v_D} = \frac{v_O/k_L}{v_{IN} + \left(v_O/k_L\right)} = \frac{v_O}{v_{IN}k_L + v_O}, \quad (58)$$

like L_I's v_{IN} for v_E and v_O/k_L for v_D and L_O's $v_{IN}k_L$ for v_E and v_O for v_D in the general expression predict.

But to return to v_O, notice that v_O is a duty-cycled scalar d_E/d_D of $v_{IN}k_L$, and together, v_O/v_{IN} scale with k_L and d_E/d_D:

$$\frac{v_O}{v_{IN}} = k_L\left(\frac{d_E}{d_D}\right) = \left(\sqrt{\frac{L_O}{L_I}}\right)\left(\frac{d_E}{1-d_E}\right). \quad (59)$$

d_E/d_D is greater than one and v_O is correspondingly greater than the transformer translation of v_{IN} when d_E is greater than 50% (and d_D is less than 50%). d_E/d_D is less than one and v_O is correspondingly less than the transformer translation of v_{IN} otherwise. So like the buck–boost, the flyback can buck *and* boost v_{IN}.

Example 15: Determine d_E when v_{IN} is 2 V, k_L is 2, and v_O is 2 V.

Solution:

$$d_E = \frac{v_D}{v_E + v_D} = \frac{v_O}{v_{IN}k_L + v_O} = \frac{2}{2(2)+2} = 33\%$$

Example 16: Determine d_E when v_{IN} is 1 V, k_L is 2, and v_O is 4 V.

Solution:

$$d_E = \frac{v_D}{v_E + v_D} = \frac{v_O}{v_{IN}k_L + v_O} = \frac{4}{1(2)+4} = 67\%$$

C. Power

Without a physical (galvanic) connection, v_{IN} cannot deliver power directly to v_O like the buck and boost can. $L_I{:}L_O$ therefore carries all the energy that v_{IN} delivers. This means that the two switches carry more current than their buck and boost counterparts. So like the buck–boost, the flyback usually burns more Ohmic power. And like the buck–boost, since v_O receives L_O's d_D fraction, the output delivers $v_Oi_{O(AVG)}$, $v_Oi_{L(AVG)}d_D$. And because v_{IN} supplies L_I's d_E fraction of the combined i_L, v_{IN} supplies $v_{IN}i_{L(AVG)}d_E$ or $v_{IN}(i_{O(AVG)}/d_D)d_E$.

D. Variants

Although ground and supply switches can (at the same time) connect and disconnect L_I from v_{IN} and L_O from v_O, only one switch per side like Figs. 25 and 27 show is necessary. A second switch would burn power, require space, and complicate the controller needlessly. Of these, the ground-switched input and supply-switched output variant in Fig. 28 is probably the most popular. Device availability, breakdown voltage, and conductivity ultimately dictate which switches are possible, more reliable, and less lossy.

E. Snubbers

In practice, parts of L_I and L_O do not couple. This means that L_O cannot drain the energy that v_{IN} injects across t_E into L_I's uncoupled fraction. So when S_{EI} opens, remnant i_{LI} charges the parasitic capacitances C_{SWI} that remain attached to S_{EI}'s switching node v_{SWI}. This is often a problem because v_{SWI} can swing above S_{EI}'s breakdown level. *Snubbers* protect switches from overvoltage conditions of this sort.

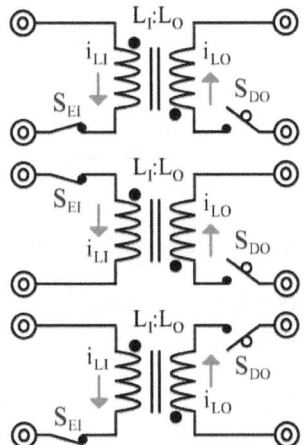

Fig. 28. Ground- and supply-switched flyback variations.

Without protection, L_I's uncoupled fraction L_I' drains into C_{SWI}, C_{SWI} drains back into L_I', and if S_{EI} does not break, C_{SWI} and L_I' exchange energy until parasitic resistances burn the energy or t_{SW} lapses. One way of limiting v_{SWI}'s swing is to dissipate some of this energy quickly. The purpose of R_{SI} in the *damper* that R_{SI} and C_{SI} implement in Fig. 29 is just this: to burn remnant energy in the core.

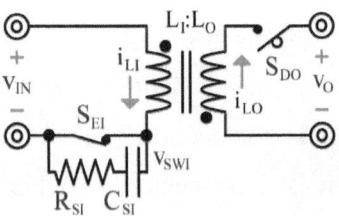

Fig. 29. Input-damped flyback.

For this, R_{SI} and C_{SI}'s combined impedance Z_{SI} at the resonant frequency f_{LC} should be lower than C_{SWI}'s Z_{SWI}. This way, Z_{SI} can steer remnant i_{LI} away from C_{SWI} into R_{SI} when S_{EI} opens. So R_{SI} burns energy, C_{SWI} peaks to a lower voltage, and L_I and C_{SI} together with C_{SWI} exchange energy across fewer cycles.

Z_{SI}, however, should not load L_I across t_{SW} to the extent that C_{SWI} cannot "fly" to $v_{IN} + v_O/k_L$. In other words, Z_{SI} should not drop to R_{SI} near or below f_{SW}. This means that R_{SI} should current-limit C_{SI} at a frequency f_{SI} that is greater than f_{SW}. But since R_{SI} should also burn LC energy, f_{SI} should also be lower than f_{LC}:

$$\left. \frac{1}{sC_{SI}} \right|_{f_{SW} < f_{SI} = \frac{1}{2\pi R_{SI}C_{SI}} < f_{LC}} = R_{SI}. \tag{60}$$

Another way to limit C_{SWI}'s swing is to clamp v_{SWI}. For this, D_S and C_S in Fig. 30 charge C_S to v_O's coupling target v_O/k_L minus D_S's diode voltage v_{DS}. If C_S is high enough to behave like a battery, C_S holds v_O/k_L $- v_{DS}$ and clamps v_{SWI} to $v_{IN} + v_O/k_L$ when S_{EI} opens. In other words, C_S absorbs remnant i_{LI} in L_I away from C_{SWI} without altering v_{SWI} too much.

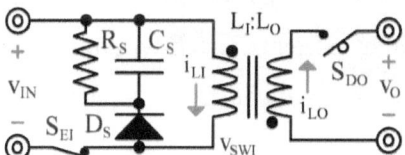

Fig. 30. Input-clamped flyback.

But since excess i_{LI} is often systemic, energy in C_S accumulates and grows over time. The purpose of R_S is to leak and burn this excess energy before the next cycle begins. If R_S is too low, however, R_S leaks energy from C_S that the core needs to replenish with what could have been delivered to v_O. Still, R_S should be lower than necessary to ensure v_{SWI} does not drift above $v_{IN} + v_O/k_L$. This additional margin dictates how much power R_S burns.

Unfortunately, the amount of L_I that does not couple depends on the manufacturing process. Plus, R_S, C_S, and D_S vary with technology, fabrication lots, and temperature. So engineers often resort to empirical methods (via simulations or experiments) when choosing R_S and C_S. Dampers are usually preferable over clampers because R_S in clampers burns not only excess L_I energy but also core energy meant for v_O. Although less frequently included, a damper across S_{DO} similarly protects S_{DO} from the effects of remnant i_{LO} in L_O.

6.2. Asynchronous Flyback

A. Power Stage

In the asynchronous flyback, L_I energizes the core with a transistor and L_O drains the core with a diode. Since electrons are more mobile than holes, an NFET burns and requires less power than a PFET under equivalent gate-drive conditions. For maximum gate drive, the source of this NFET should connect to the lowest potential. This is why M_{EI} in Fig. 31 is the N-channel ground switch that energizes L_I. The source points to v_{IN}'s ground (away from the channel through which i_{LI} flows) because N-channel sources output current. The P-type bulk also connects to v_{IN}'s ground to keep M_{EI}'s body diode from conducting i_{IN} when v_{LI} "flies" high.

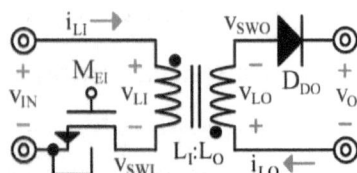

Fig. 31. Asynchronous flyback.

M_{EI} energizes the core by feeding i_{LI} into L_I's dotted terminal. L_O must therefore drain i_{LO} out of the opposite (non-dotted) terminal. This is why D_{DO} in Fig. 31 is the supply switch that drains i_{LO} into v_O. Since D_{DO} blocks reverse current, v_{LO} "flies" high when M_{EI} energizes L_I. Although connecting D_{DO} in v_O's ground path also works, grounding L_O without a switch in series ties L_O to the least noisy terminal with lower resistance.

B. Duty-Cycle Translation

Since D_{DO} drops a diode voltage above v_O, L_O drains the core with v_O plus D_{DO}'s diode voltage v_{DO}. This higher v_D drains the core faster than in the ideal flyback, so t_D and t_C are shorter and t_E is a larger fraction of t_C. In other words, d_E is higher than in the ideal flyback.

v_{LI} therefore swings between v_{IN} and $-(v_O + v_{DO})/k_L$ and v_{LO} swings between $v_{IN}k_L$ and $-(v_O + v_{DO})$. In other words, v_{LI} is v_{IN} a d_E' fraction of t_C, $-(v_O + v_{DO})/k_L$ a d_D' fraction, and zero on average:

$$v_{LI(AVG)} = v_{IN}d_E' - \left(\frac{v_O + v_{DO}}{k_L}\right)d_D'$$

$$= v_{IN}d_E' - (v_O + v_{DO})\left(\sqrt{\frac{L_I}{L_O}}\right)d_D' = 0. \tag{61}$$

v_{LO} is similarly $v_{IN}k_L$ a d_E' fraction of t_C, $-(v_O + v_{DO})$ a d_D' fraction, and zero on average:

$$v_{LO(AVG)} = v_{IN}k_Ld_E' - (v_O + v_{DO})d_D'$$

$$= v_{IN}\left(\sqrt{\frac{L_O}{L_I}}\right)d_E' - (v_O + v_{DO})d_D' = 0 \tag{62}$$

Since d_D' is $1 - d_E'$, the diode raises L_O's v_D or v_O in the numerator of d_E' by a larger fraction than in the denominator that v_E's $v_{IN}k_L$ and v_D's v_O set:

$$d_E' = \frac{v_D}{v_E + v_D} = \frac{(v_O + v_{DO})/k_L}{v_{IN} + (v_O + v_{DO})/k_L} = \frac{v_O + v_{DO}}{v_{IN}k_L + v_O + v_{DO}} > d_E. \tag{63}$$

So the effect of the diode is to increase d_E.

Example 17: Determine d_E' when v_{IN} is 1 V, k_L is 2, v_O is 4 V, and v_{DO} drops 800 mV.

Solution:

$$d_E' = \frac{v_O + v_{DO}}{v_{IN}k_L + v_O + v_{DO}} = \frac{4 + 800m}{(1)(2) + 4 + 800m} = 71\%$$

Note: d_E' here is higher than d_E in the ideal example because D_{DO} increases L_O's drain voltage. So t_D shortens and t_E's fraction of t_C's $t_D + t_E$ increases. d_E's variation (from ideal to asynchronous) is higher than in the buck–boost because only one diode raises v_D – two diodes raise v_D in the buck–boost.

C. Conduction Modes

In continuous conduction, the coupled inductor pair conducts a combined current i_L or $i_{LI} + i_{LO}$ about an average $i_{L(AVG)}$ that keeps i_L above zero. When transferring little power, $i_{L(AVG)}$ can be so low that i_L reaches zero before t_{SW} lapses, like the thicker traces in Fig. 32 illustrate. Since D_{DO} cannot conduct reverse current, i_L remains near zero until the next t_{SW}. $L_I{:}L_O$ is in discontinuous conduction when this happens.

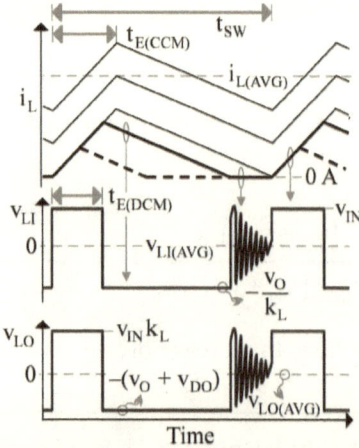

Fig. 32. Discontinuous-conduction waveforms in the flyback.

When L_I is drained and disconnected, v_{IN} and v_{SWI}'s drained parasitic capacitance C_{SWI} impress a voltage across L_I that draws power from the input and energizes L_I and C_{SWI}. i_{LI} begins to drain L_I into C_{SWI} after v_{SWI} rises above v_{IN}. When L_I depletes, v_{SWI} and v_{IN} impress a voltage across L_I that drains C_{SWI} into L_I and v_{IN}. L_I then drains into v_{IN} when v_{SWI} falls below v_{IN}. v_{SWI} is again below v_{IN} when L_I depletes, so the entire sequence repeats. v_{LI} oscillates with v_{SWI} this way about zero until resistances burn the energy or t_{SW} lapses.

v_O and D_{DO}'s v_{SWO} similarly impress a voltage across L_O when the core depletes that drains v_{SWO}'s parasitic capacitance C_{SWO} into L_O. L_O then drains into C_{SWO} to charge C_{SWO} in the negative direction, C_{SWO} drains back into L_O, L_O depletes into C_{SWO}, and the sequence repeats. v_{LO} oscillates with v_{SWO} about zero until resistances burn the energy or t_{SW} lapses.

6.3. Synchronous Flyback

A. Power Stage

The difference between asynchronous and synchronous power supplies is what drains the core: diodes in one and transistors in the other. So the ground input FET M_{EI} that energizes the asynchronous flyback and implements S_{EI} in the ideal flyback also energizes the synchronous counterpart in Fig. 33. The difference here is that the output transistor M_{DO} drains L_O, which is the function of S_{DO} and D_{DO} in the ideal and asynchronous flybacks.

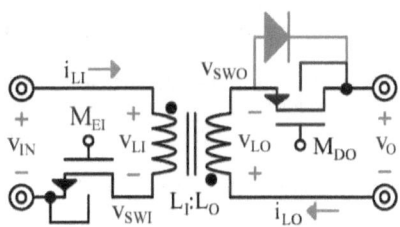

Fig. 33. Synchronous flyback.

When closed, M_{DO}'s source and drain voltages v_S and v_D are close to v_O. M_{DO}'s gate voltage v_G can swing reliably between v_O and ground. With these voltages, the maximum gate drive of an NFET would be v_{GST}, $v_G - v_S - v_{TN}$, $v_O - v_O - v_{TN}$, or just $-v_{TN}$, which is too low to close an NFET. The maximum gate drive of a PFET would be v_{SGT}, $v_S - v_G - |v_{TP}|$, $v_O - 0 - |v_{TP}|$, or $v_O - |v_{TP}|$, which is greater and more likely to close the switch. This is why M_{DO} is oftentimes a PFET, because a high v_S calls for a high-side switch.

Since PMOS source terminals receive current, L_O feeds M_{DO}'s source and M_{DO}'s source arrow points into the transistor (into the channel through which i_L flows). The N-type bulk terminal connects to v_O to block reverse i_O when v_{SWO} falls to $-v_{INK_L}$. This connection also allows the remaining body diode to steer L_O's dead-time current into v_O.

B. Duty-Cycle Translation

The difference between asynchronous and synchronous operation across t_D is that M_{DO} drops a lower voltage than D_{DO} does. So v_{SWO} is close to v_O when M_{DO} closes within t_D. But since M_{EI} and M_{DO} open across the dead-time portions of t_D and M_{DO}'s body diode offers a drain path for i_L, v_{SWO} reaches $v_O + v_{DO}$ across the t_{DT}'s that t_D incorporates.

This means that L_O drains two t_{DT} fractions of t_C with v_O plus a diode voltage and the remainder of t_D's fraction with v_O. This higher v_D drains the core faster than in the ideal flyback, but not as fast as in the asynchronous flyback because v_D is higher only momentarily. t_D and t_C are therefore shorter, but not as short as in the asynchronous flyback. t_E is similarly a longer fraction of t_C, but also not as long a fraction as the asynchronous d_E' is.

v_{LI} is therefore v_{IN} a t_E fraction of t_C, $-(v_O + v_{DO})/k_L$ two t_{DT} fractions, $-v_O/k_L$ the fraction of t_D that remains, and zero on average

$$v_{LI(AVG)} = v_{IN}d_E'' - \left(\frac{v_O + v_{DO}}{k_L}\right)\left(\frac{2t_{DT}}{t_C}\right) - \frac{v_O}{k_L}\left(\frac{t_D - 2t_{DT}}{t_C}\right) = 0. \qquad (64)$$

v_{LO} is similarly $v_{IN}k_L$ a t_E fraction of t_C, $-(v_O + v_{DO})$ two t_{DT} fractions, $-v_O$ the fraction of t_D that remains, and zero on average:

$$v_{LO(AVG)} = v_{IN}k_Ld_E'' - (v_O + v_{DO})\left(\frac{2t_{DT}}{t_C}\right) - v_O\left(\frac{t_D - 2t_{DT}}{t_C}\right) = 0. \qquad (65)$$

Since d_D'' is $1 - d_E''$, the diode raises v_D or v_O in the numerator of d_E' by a larger fraction than in the denominator that v_E's $v_{IN}k_L$ and v_D's v_O set:

$$d_E'' = \frac{v_O/k_L}{v_{IN} + (v_O/k_L)} + \left[\frac{v_{DO}/k_L}{v_{IN} + (v_O/k_L)}\right]\left(\frac{2t_{DT}}{t_C}\right)$$

$$= \frac{v_O}{v_{IN}k_L + v_O} + \left(\frac{v_{DO}}{v_{IN}k_L + v_O}\right)\left(\frac{2t_{DT}}{t_C}\right)$$

$$= d_E + \left(\frac{v_{DO}}{v_{IN}k_L + v_O}\right)\left(\frac{2t_{DT}}{t_C}\right) > d_E. \qquad (66)$$

Example 18: Determine d_E'' when v_{IN} is 1 V, k_L is 2, v_O is 4 V, v_{DO} drops 0.7 V, t_{DT} is 50 ns, and t_C is 1 µs.

Solution:

$$d_E'' = \frac{v_O}{v_{IN}k_L + v_O} + \left(\frac{v_{DO}}{v_{IN}k_L + v_O}\right)\left(\frac{2t_{DT}}{t_C}\right)$$

$$= \frac{4}{1(2) + 4} + \left[\frac{0.7}{1(2) + 4}\right]\left[\frac{2(50n)}{1\mu}\right] = 68\%$$

Note: d_E'' here is higher than d_E in the ideal example because M_{DO}'s body diode raises v_D. But since the diode does so only two t_{DT} fractions of t_C, t_D shortens and t_E's fraction of t_C rises less than d_E' in the asynchronous example.

C. Conduction Modes

If the controller opens and closes M_{DO} when the asynchronous diode D_{DO} would, the only difference between asynchronous and synchronous operation is the voltage dropped across the switch: millivolts with FETs and 0.60–0.75 V with diodes. But since M_{DO} is off across dead-time periods, M_{DO}'s body diode conducts across t_{DT}'s like D_{DO} in the asynchronous flyback.

If the controller does not open M_{DO} when i_L falls to zero before t_{SW} lapses, M_{DO} lets i_L reverse direction. Since M_{DO}'s body diode cannot conduct this negative i_L across the t_{DT} that follows, M_{EI}'s body diode conducts this i_L into v_{IN}. So v_{SWI} falls a diode voltage across this t_{DT} and returns to zero when M_{EI} closes. Returning energy to v_{IN} this way means L_X transfers and burns more power than necessary.

Note one t_{DT} is in t_D and another in t_E when i_L reverses, whereas without negative conduction, t_D includes both t_{DT}'s. And since v_{SWI} falls and v_{SWO} rises by a similar diode voltage across similar t_{DT}'s, diode effects on $v_{L(AVG)}$'s tend to cancel. So with negative conduction, d_E is close to the ideal case (lower than d_E").

D. Diode Conduction

If M_{DO}'s threshold voltage is lower than a diode voltage, M_{DO} closes across t_{DT}'s when v_{SWO} climbs a $|v_{TP}|$ above M_{DO}'s v_O-supplied gate voltage. M_{DO}'s body diode does not inject noise current into the substrate when this happens. A Schottky diode across M_{DO} similarly steals dead-time current away from the body diode and the substrate into which the parasitic BJT present injects current.

7. Summary

Inductors manifest magnetic energy in the form of current. And inductors that share magnetic space can energize and drain that space with voltages of opposing polarities. This way, switched inductors and transformers transfer input power to output loads. Series resistances, however, burn some of this energy. Plus, nearby coils and fast-changing currents restrict current flow. So series resistances climb with more nearby coils and higher switching frequency.

Many consumer products transfer power this way from ac–dc chargers and internal batteries to electronic systems that require stable dc power supplies. The inputs and outputs of the switched inductor are therefore static or quasi-static dc voltages. These dc voltages ramp up and down the inductor's current. Since inductors receive and output similar energy levels, these same voltages set the energizing and draining duty-cycle fractions of the switching period.

To supply the energy that series resistances burn, switched inductors must energize a larger duty-cycle fraction. This way, they can overcome losses and deliver power continuously or discontinuously. For this, CMOS solutions use MOSFETs to energize inductors and diodes or MOSFETs to drain them. Available gate drive and dead-time current-flow requirements dictate which type of MOSFET is optimal for each switch in the network.

With four switches, switched inductors can buck *and* boost input voltages to lower or higher output voltages. Removing the two input or two output switches sets an average voltage that only a higher voltage at the opposite end can establish. This way, two switches can buck *or* boost, but not both.

Asynchronous circuits drain the inductor with diodes and synchronous circuits with MOSFETs. To avoid momentary shorts, synchronous solutions include dead time between the conduction periods of adjacent switches. Body, MOS, or Schottky diodes conduct the inductor's current across these times.

In flybacks, an input inductor magnetizes the space that an output inductor drains. This way, input sources and output loads need not share a common ground (for galvanic isolation). But since parts of the input inductor do not couple to the output, engineers use snubbers to burn leftover energy.

Overall, switched inductors are vital in microelectronics. For one, they can boost input voltages to higher output levels, whereas linear transistors cannot. Plus, even when bucking, they burn less power than their non-switched counterparts. So engineers often rely on them to charge batteries and establish regulated power supplies with minimal power losses.